# Paving the Way to a Greener Future

## *Eco-Friendly Interlocking Pavers*
## *for Sustainable Landscapes*

Arreshvhina Narayanan

# Copyright Page



First Edition, 2025
Published by Notion Press (India. Singapore. Malaysia)
Printed in India

For permissions, licensing, or bulk orders, contact:
arreshvhina@gmail.com

## Synopsis

What if a simple paver could help cool a city, recharge the ground, and reduce carbon all at once?

In *Paving the Way to a Greener Future*, civil engineer and sustainability strategist Arreshvhina unpacks the overlooked power of one of the most common elements in the built environment: the interlocking concrete paver. This is not just a construction manual. It is a systems-level guide with in-depth technical, practical, and visionary insights for professionals who want to transform paving from a passive surface into a proactive climate solution.

Drawing on 17+ years of field and manufacturing experience, the book covers:

- The science behind permeability, reflectivity, algae resistance, and durability
- Low-carbon material innovation, recycled aggregates, and circular production
- Smart design for subbase performance, surface patterning, drainage, and disassembly
- Failure modes and lessons learned from real-world case applications
- Lifecycle costing, Environmental Product Declarations (EPDs), and maintenance planning
- How paving systems can support larger urban, ecological, and infrastructural goals

Whether you're an engineer, architect, planner, contractor, or policymaker, this book challenges you to rethink how every paved surface can become a resilient, regenerative asset, and not just something you walk on.

Because sustainability doesn't start at the rooftop or the skyline.
It starts under your feet.

# Preface

This book was born out of 17+ years of experience working in the field of concrete manufacturing, materials innovation, and sustainability-driven infrastructure design. For most of my career, interlocking pavers were seen as a basic civil item; always measured by strength, colour, and cost. But as the climate changed, so did the questions.

What if pavements could cool cities? What if they could recharge aquifers? What if they could be reused, reimagined, and recycled instead of ripped up and dumped?

This book is my response to those questions. It's not a catalogue of products; it's a knowledge exploration of what paving *can* be when it's treated not as an afterthought, but as a strategic tool. You'll find practical detail here, yes. But more importantly, I hope you'll find a shift in mindset: from installing surfaces to designing outcomes.

Whether you're a civil engineer, urban planner, architect, developer, contractor, or policy thinker, this guide is for you.

Let's stop paving over the problem. Let's start paving the way forward.

- Arreshvhina -

# Table of Content

# Chapter 1: Concrete Meets Conscience

*Why Pavers Matter in a Warming World*

## 1.1 The Ubiquity and Problem of Conventional Surfaces

Concrete is one of the most widely used materials on Earth, second only to water in sheer volume. From car parks and courtyards to sidewalks and stormwater channels, it forms the backbone of our built environment. It's reliable, affordable, and easy to manufacture. But in many outdoor applications, particularly urban hardscaping, its dominance has become problematic.

We've built for decades on a simple assumption: if it's strong and solid, it works. But the world around that assumption has changed; climate patterns have shifted, cities have densified, and expectations around sustainability have evolved. The cracks are no longer just in the pavement, they're in the way we think about the role of paving itself.

## 1.2 The Hidden Cost of Heat

Traditional concrete surfaces absorb and store solar radiation. During the day, they heat up rapidly, and at night, they release that heat back into the environment. This contributes to the urban heat island (UHI) effect, where cities become significantly warmer than nearby rural areas due to the prevalence of impervious, heat-retaining materials.

While specific intensity varies by geography, UHI has been consistently documented in cities worldwide. According to research published in *Nature Communications* (2022), some urban centres can experience surface temperature increases of up to 10°C compared to adjacent rural land.

This increase drives higher energy consumption, intensifies heatwaves, and creates harsh microclimates; especially in low-vegetation areas where hardscape dominates.

*"A cooler city doesn't start with more trees,*
*It starts with fewer heat traps beneath them."*

## 1.3 When Water Has Nowhere to Go

Another downside of conventional slabbed surfaces is poor water permeability. Instead of soaking into the soil, rainwater collects on the surface and rapidly channels into stormwater systems, often overwhelming them. In dense cities or industrial parks, this runoff not only contributes to flash flooding, but also picks up contaminants such as oil, sediment, litter; before it flows untreated into rivers and seas.

In regions with monsoon climates, such as Southeast Asia, where intense rain can occur in short bursts, the implications are even more severe. Surfaces that cannot absorb water become liabilities for both in environmental and economic terms.

*"If the ground beneath our feet can't breathe or drink,*
*the city around it will drown or overheat.*
*Concrete control has become climate control."*

## 1.4 A Maintenance Headache

Concrete slabs don't just fail under stress, they become harder to live with over time. Algae buildup in humid, shaded areas creates slip hazards. Cleaning requires water, chemicals, and labour. Cracks or structural failures necessitate large-area removal and full slab replacement.

This kind of surface lifecycle is expensive, disruptive, and far from sustainable. What's worse, poor sub-base design often accelerates deterioration, especially under repeated loading and exposure cycles. And yet, conventional designs continue to persist, not because they perform better, but because they are familiar.

*"Slabs crack, algae creeps, drains clog—and the budget bleeds. But we keep pouring concrete like it's still the 1980s."*

## 1.5 A Clearer Comparison

To move the conversation forward, we need a clearer picture. Below is a side-by-side view of how traditional slabs stack up against interlocking concrete pavers when measured through a sustainability lens:

| Aspect | Traditional Slabs | Interlocking Pavers |
| --- | --- | --- |
| Water infiltration | Impervious surface | Permeable or semi-permeable (joint or porous) |
| Thermal performance | Low SRI (<30), absorbs heat | High SRI (>50 possible), reflects sunlight |
| Algae control | High in shaded/humid areas | Algae-resistant formulations reduce maintenance |
| Maintenance | Requires large-area removal | Modular—single unit replacement possible |
| Aesthetic flexibility | Limited texture and tone options | High variability in pattern, texture, and color |
| Lifecycle emissions | Higher due to rigid construction cycles | Lower if manufactured with eco-binders/recycled content |

## 1.6 Engineering Rigidity vs. Resilience

Concrete has always symbolized permanence. But in a time of environmental variability, permanence isn't always an advantage. Urban surfaces now must deal with expansion and contraction from heat, sudden rain events, longer dry spells, and unpredictable load scenarios. The ability to *adapt* is to be modular, replaceable, water-sensitive, and reflective; and these factors has become as important as the ability to resist compressive load.

*"Resilience is no longer optional in engineering. It's the benchmark."*


## 1.7 Why This Book Starts with the Surface

The ground beneath us has always been functional. Now it needs to be intelligent. Reflective, permeable, modular, algae-resistant; these aren't buzzwords. They're performance traits that align with where cities are going.

In the chapters ahead, we'll explore:
- What makes paving truly sustainable
- How to design it, specify it, install it, and maintain it
- And how to convince clients, consultants, contractors; that the smarter surface is worth the switch

Because the future isn't poured.
It's placed one unit at a time.

# Chapter 2: What Makes Paving Eco-Friendly

*Beyond Buzzwords: The Engineering of Sustainable Surfaces*

For all the space that "sustainability" occupies in planning documents, procurement pitches, and government targets, the term often loses its weight when it hits the ground. In the world of paving, "eco-friendly" is thrown around like a sticker: something to slap on any product that isn't overtly harmful.

It's easy to blame marketing. Many material suppliers inflate environmental claims with phrases like *low-impact, green-enhanced*, or *rain-ready*; words that sound good but say little. But the real issue runs deeper. Even among professionals e.g. engineers, architects, municipal planners; there's often no shared technical understanding of what "eco-friendly paving" really means in practice.

The result?
- Specifications that list "permeable pavers" without defining infiltration rates
- Material choices made for colour or availability, not climate performance
- Projects that tick sustainability boxes without achieving meaningful outcomes

**"Eco-friendly" is not a feature. It's a systemic quality that emerges only when design, material, context, and performance are aligned.**

Let's question a few lazy assumptions:
- Is a permeable paver always sustainable? *(No if it clogs in two months.)*
- Are recycled materials always better? *(No if they compromise strength or leach toxins.)*
- Does high SRI automatically mean cooler cities? *(Not without placement, orientation, and maintenance.)*

***Validation:** True sustainability is measurable. It shows up in infiltration rates, surface temperatures, cleaning cycles, VOC content, embodied carbon, modularity, and lifespan, and not in vague promises.*

### What actually makes a paving system eco-friendly?

It's not just about swapping one material for another. It's not just about green-coloured products or porous-looking bricks.

A truly eco-friendly paving system considers:
- **Water** – how the surface interacts with rainfall, drainage, and groundwater
- **Heat** – how it absorbs, reflects, or radiates solar energy
- **People** – how it supports safe movement, comfort, and maintenance
- **Time** – how it performs not just on day one, but over decades

> **"A sustainable surface doesn't just solve today's problems,**
> **It anticipates tomorrow."**

This chapter breaks that down, section by section. We'll explore what performance looks like in the field and not just in product brochures, and how to specify paving systems that deliver measurable environmental value across climate zones, site types, and use cases.


## 2.1 Permeability and Water Sensitivity
*Why paving systems must behave like landscapes—not lids.*

Rain is predictable. What's not predictable is how well we've prepared our built environment to deal with it. In cities across the world; from Amsterdam to Kuala Lumpur, Vancouver to Lagos, stormwater management is among the most pressing issue, yet often most poorly addressed challenges in urban infrastructure.

We design streets around cars, regulate buildings for daylight, and zone districts for population density. But water arguably the most volatile and system-shaking element gets handled late in the design cycle, and too often, poorly.

Traditional paving surfaces such as impervious slabs, monolithic concrete, and tightly bound asphalt be likely to break the natural hydrological cycle. They interrupt infiltration, accelerate runoff, and act as heat-absorbing, pollutant-spreading surfaces. Rain, which once filtered quietly into the ground, now becomes a fast-moving, high-volume discharge event. It picks up hydrocarbons, heavy metals, debris, and microplastics. It overwhelms storm drains, floods basements, and scours unprotected perimeters. It doesn't just "go away"; it becomes someone else's problem downstream.

In this context, permeable paving is not just a sustainable alternative, it's a system corrective. It allows surfaces to function more like soil: absorbing, slowing, and filtering water in place. When properly designed, these systems reduce peak flow, lower runoff volumes, improve water quality, and recharge groundwater—thereby extending the lifespan and reducing the load on municipal drainage infrastructure. But the key word here is "designed." Because permeable paving doesn't just mean punching holes in a surface; it requires a complete shift in how the system is conceptualized and built, from surface to subgrade.

## Understanding the Types of Permeable Systems

There are three primary categories in global practice: Jointed Permeable Pavers (JPPs), Permeable Concrete Pavers (PCPs), and hybrid or composite systems that integrate aspects of both.

**JPPs** are perhaps the most common in both residential and light commercial applications. These systems consist of traditional interlocking concrete units, spaced apart with engineered joints which typically 5 to 15 mm wide then filled with clean, angular aggregates.

The surface functions by directing water vertically through these voids into a layered, permeable substructure. The units themselves remain dense and non-porous, which allows for superior compressive strength and dimensional tolerance. Their modularity makes them easy to maintain: if a joint fails or clogs, only local replacement is needed.

**PCPs**, on the other hand, use specially formulated low-fines concrete to create internal porosity within the paver unit itself—think of them as sponge bricks. Water passes through the paver body directly, entering the bedding and base layers underneath. These systems tend to provide higher surface permeability per square meter and are ideal in plazas or areas with consistent low to medium traffic. However, their open matrix is more sensitive to fines intrusion, wear, and structural stress, especially under repetitive vehicular loads or freeze-thaw cycles in temperate climates.

In practice, **hybrid systems** often emerge: PCPs installed with engineered joint gaps, or JPPs paired with vegetated strips or underground retention modules. The design choice depends heavily on intended use, rainfall intensity, maintenance expectations, and subgrade conditions.

⚠ *A common design pitfall: engineers often specify permeable pavers while reusing subbase designs intended for slab-on-grade concrete. This creates a disconnect where infiltration is possible at the surface, but impossible below.*

## Subbase Engineering: The Hidden Success Factor

If the surface is the face of a permeable system, the ***subbase is the brain and lungs.*** It dictates whether infiltration is retained, distributed, or silently compromised. Subbase design must not only hold weight, it must do so while maintaining voids, resisting compaction under saturation, and draining predictably.

A typical high-performance system will begin with a ***filter layer*** or nonwoven geotextile, separating the subgrade soil from the subbase to prevent fines migration. Above this lies the ***reservoir layer***, often composed of 30–70 cm clean, angular crushed stone with a void content of 35–45%. This layer stores stormwater during heavy rainfall events—offering detention capacity while infiltration occurs over time.

A base course, usually 20–40 cm aggregate, supports load transfer and interlock. The bedding layer, typically 2–5 cm stone or washed grit, ensures levelling and helps anchor joint fill.

Depending on local soil infiltration capacity, systems may be designed as:
- **Full infiltration** (native soil accepts stormwater)
- **Partial infiltration with underdrain** (clay soils, high water table)
- **No infiltration (detention only)**, where water is conveyed to adjacent stormwater infrastructure

Standards and guidelines from ICPI (Interlocking Concrete Pavement Institute), BS 7533, and ASTM C1781 provide methodologies for infiltration testing, base layer compaction, and drainage integration. In regions like the UK and Netherlands, CEEQUAL and SuDS frameworks explicitly encourage such systems, linking their adoption to stormwater credits and planning approvals.


## Why Permeable Systems Matter—Locally and Globally

Urban flooding is now a global phenomenon. Whether it's monsoonal flash floods in Jakarta, sewer overflows in New York, or non-point source pollution in Brisbane, the failure of surface water management has consequences far beyond individual sites. And the opportunity for permeable paving isn't just in new development, it lies in retrofitting and decentralizing stormwater solutions across aging city grids.

When paired with vegetated bioswales, tree pits, green roofs, or rain gardens, permeable surfaces become nodes in a distributed water-sensitive urban design (WSUD) network. This aligns with urban climate adaptation policies under the UN-Habitat guidelines, the Global Covenant of Mayors, and ISO 37123:2019 (*Indicators for Resilient Cities*), all of which emphasize nature-based stormwater interventions.

Beyond flooding, permeable paving supports broader ecological goals:
- **Reduces thermal mass**, lowering surface temperatures
- **Minimizes algae-friendly pooling**, especially in shaded areas
- **Enhances aquifer recharge**, helping stabilize local water tables
- **Filters first-flush runoff**, removing zinc, oil, rubber dust, and heavy metals before they enter natural systems

*"Sustainability starts when your surface stops acting like a barrier, and starts behaving like soil."*

And in that soil-like behaviour lies its power; not only as a niche solution, but as a mainstream material logic for climate-resilient urbanism.


## 2.2 Solar Reflectance Index and Heat Management
*Why a surface's temperature is more than skin deep.*

In dense cities, paving materials don't just sit underfoot; they act as massive thermal batteries. Every square metre of exposed hardscape collects solar radiation during the day and re-releases it slowly through the night, contributing to what is now globally recognized as the ***Urban Heat Island (UHI) effect***. The hotter a surface gets, the more it radiates warmth back into the surrounding air. And unlike vertical surfaces that can be shaded or ventilated, pavements are constantly exposed; thus, constantly absorbing sunlight, storing heat, and amplifying discomfort at street level.

Here, material choice becomes not just a matter of aesthetics or durability, but a climate mitigation strategy. The Solar Reflectance Index (SRI) offers a way to quantify how much a surface contributes to or alleviates this effect.

It's a composite metric, combining two physical properties:
- **Albedo**, or solar reflectance — how much sunlight is reflected away from the surface
- **Thermal emissivity** — how efficiently a material radiates absorbed heat back into the atmosphere

These are measured under standardized conditions (ASTM E1980 being the most widely accepted test protocol), with SRI values ranging from 0 (black, heat-absorbing surfaces like fresh asphalt) to 100 (idealized white, high-emissivity surfaces). Most common materials fall somewhere between:

| Material | Typical SRI |
| --- | --- |
| Black asphalt | 0–10 |
| Standard grey concrete | 25–35 |
| White concrete | 70–80 |
| High-SRI engineered pavers | 50–75+ (with additives) |

The implications of these numbers are not academic. They affect real energy use, real health outcomes, and real material lifespans in the built environment.

## Why SRI Isn't Just a Colour or Finish Decision

It's tempting to see SRI as a design afterthought—a choice between grey and beige, or whether a light finish fits the site's visual tone. But high-SRI paving plays a system-level role in urban performance. Its effects are structural, thermal, and behavioural.

In terms of urban heat mitigation, studies by NASA, the U.S. EPA, and Singapore's Urban Redevelopment Authority have shown that high-SRI surfaces can reduce pavement temperatures by 8°C to 20°C depending on region, time of day, and surface texture. In high-density tropical cities like Bangkok or Manila, this doesn't just improve walkability; it lowers mortality risk during heatwaves.

Beyond ambient air temperatures, high-SRI pavers influence **building energy loads**. Pavement near building facades reflects radiant heat back into vertical surfaces, particularly glass and metal. This **raises interior cooling loads,** especially in single-story structures, retail strips, and campuses. By lowering surface temperatures, SRI-optimized materials can reduce these indirect heat gains—producing tangible energy savings, particularly where passive cooling or green envelope strategies are used.

There's also the material science angle: when surfaces heat up and cool down rapidly, particularly in climates with high diurnal ranges, they undergo repeated expansion and contraction. Over time, this leads to:

- Joint instability
- Cracking or surface delamination
- Early fatigue and accelerated aging

High-SRI materials experience smaller thermal gradients, meaning less stress, fewer repairs, and better long-term dimensional stability.

*In this context, SRI isn't just a surface spec—it's a long-term stress reducer, both for people and the pavement itself.*

# SRI in Standards, Certification, and Urban Policy

Because of its broad influence, SRI is now embedded in most green certification frameworks and infrastructure rating tools. For instance:

1. **LEED v4.1 (USGBC)**
   - Credit: *Heat Island Reduction*
   - Paving materials must achieve **SRI ≥ 29**
   - Credit multipliers exist for using high-SRI materials without vegetation
   - Measurement protocol: ASTM E1980, ASTM C1549 (reflectance), and ASTM C1371 (emissivity)

2. **GreenRE & MyHIJAU (Malaysia)**
   - Encourage passive cooling strategies through surface selection
   - Include SRI benchmarks in the evaluation of *non-roof hardscape* components
   - Integrated into broader "Cool Urbanism" and "Climate-Resilient Infrastructure" tracks

3. **SITES & Envision (Landscape and Infrastructure)**
   - Use SRI to evaluate human comfort, urban climate impact, and lifecycle performance
   - Allow integration of data into site-wide thermal analysis models

In regions with mandatory climate action plans, such as the EU Green Deal or California's CALGreen; local governments are increasingly mandating high-albedo materials in new developments and retrofits, especially for roads, pedestrian plazas, and open-air public amenities.

## SRI Over Time: Degradation and Design Considerations

One of the most critical misunderstandings around SRI is that it's a static property. In reality, most high-SRI surfaces begin to lose reflectivity over time due to:

- **Surface staining from dust, oil, or biological growth**
- **Abrasion from foot or wheel traffic**, especially where loose aggregates or sand are present
- **Degradation of coatings**, sealers, or unstable pigments from UV exposure

That's why engineers and specifiers should avoid relying on post-application coatings to achieve SRI compliance. Instead, prefer integrally pigmented surfaces, naturally reflective aggregates, and hydrophobic treatments that discourage biological buildup. In humid or shaded environments, pairing SRI goals with algae-resistant surface technology ensures that light-coloured pavements don't become safety hazards within a year.

Field studies in region with high temperature, have shown that SRI can drop by 10–30% within the first two years if materials aren't properly stabilized. Conversely, pavers made with white quartz, limestone fines, or light-toned GGBS blends retain performance better and require less maintenance.

## Engineering the Thermal Surface

So what does good thermal paving design look like in practice?

It starts with material selection; but continues with layout orientation, adjacency planning, and climate-resilient detailing. Light-coloured pavers should be deployed most intensively in pedestrian plazas, play areas, transit zones, and low-vegetation sites. In zones where greenery is possible, SRI can complement; but can never replace the vegetative shading. And in areas with strong glare risk (e.g., near glass storefronts), reflectivity should be balanced against visual comfort.

In short: SRI is one variable in a larger microclimate equation. When used correctly, it multiplies the effect of other interventions like cool roofs, urban trees, passive ventilation, and even energy grid performance.

## 2.3 Algae-Resistant Surfaces

*Why clean design isn't just visual—it's functional, financial, and safer.*

In most northern climates, concrete degrades from the top down—frost, de-icing salts, UV exposure. But in the tropics and subtropics, the degradation often begins not with cracking, but with a slippery green film. In shaded courtyards, under awnings, near landscaped planters, or wherever water lingers, algae find a foothold, and once it appears, it rarely retreats quietly.

This isn't a cosmetic concern. It's a safety and maintenance liability, especially in regions with high rainfall, persistent humidity, and dense tree cover. Moisture doesn't evaporate easily, surfaces stay wet longer, and organic debris settles fast. The result is a perfect bio-environment for algae, mold, and mildew, often within months of installation.

On untreated concrete or light-coloured pavers, the sequence is predictable: first dark streaks, then surface slickness, then frequent complaints. Cleaning crews are dispatched weekly with pressure washers, detergents, and sometimes acids. Tenants complain. Facility managers budget for extra manpower. And if someone falls, especially a child or elderly person, the costs escalate quickly - from surface rehab to potential legal action.

What's often missed is that this entire scenario is not a cleaning problem but a design problem. Therefore, like any design problem, it can be mitigated or avoided entirely with the right surface strategy.

## The Material Science Behind Algae Resistance

Effective algae control begins not at the maintenance plan, but at the mix design and surface finish stage. Algae growth is influenced by three factors: moisture availability, surface roughness, and chemical environment. Modern algae-resistant paving solutions target all three.

First, micro-structured surfaces disrupt the water film that algae depend on. These surfaces are not just rough—they are textured at the micron level, reducing the surface tension that allows water to pool. Inspired by biomimetic principles (e.g., lotus leaves, shark skin), these surfaces promote rapid drying after rain or dew. Faster drying means shorter wet periods—reducing the window for algal spores to adhere, colonize, and thrive.

Second, pH-modified binders chemically alter the environment. Algae prefers mildly acidic to neutral conditions (pH 6–7.5). Many modern paver formulations intentionally raise surface pH using supplementary cementitious materials (SCMs) such as GGBS or fly ash, or with tailored admixture packages that sustain a high pH (~10–12) near the surface even after carbonation. These conditions are not permanently biocidal, but they discourage microbial settlement and retard growth cycles.

Finally, hydrophobic surface treatments; not the sealants, but thin, breathable coatings which will further reduce water absorption and organic accumulation. These treatments create a low-energy surface that repels moisture without trapping it underneath, allowing moisture vapor to escape while minimizing the surface's attraction to dust, spores, and pollution.

*"It's not about stopping algae from existing; it's about making the surface a place where algae can't gain a foothold."*

## Performance, Cost, and Risk Reduction

When integrated properly, algae-resistant paving delivers clear, long-term benefits across multiple operational categories:

- **Lower water usage**—Weekly pressure washing in tropical sites consumes thousands of litres per cycle. A paver that stays dry cuts this dramatically.
- **Less chemical runoff**—Detergents, biocides, and degreasers often end up in storm drains, contradicting broader environmental goals.
- **Longer cleaning intervals**—Surfaces that resist algae require intervention monthly or seasonally, not weekly; thus, reducing labour and noise pollution.
- **Improved surface safety**—Especially in hospital access roads, school zones, pool decks, or transit terminals where vulnerable populations walk.
- **Longer surface lifespan**—Repeated power washing degrades pigments, erodes micro-texture, and prematurely ages sealers. Cleaner surfaces mean gentler care.

In facilities where incident rates are monitored, algae-prone surfaces are often linked to a higher occurrence of slips and falls—a leading cause of liability claims in retail, healthcare, and hospitality sectors.

*"You're not just keeping the surface clean…*
*You're keeping people safe; budgets lean and drains clear."*

## Design Considerations for Tropical and Urban Environments

Algae resistance must be treated as a primary design criterion, not a post-failure fix. This means asking the right questions at the specification stage:

- Does the product use integral pH-control in its binder composition?
- Is the surface finish hydrophobic and breathable, not film-forming?
- Does the texture promote drainage and rapid drying, or does it trap water?

- Has the product been tested under biological colonization protocols such as ASTM D5590 or in regional climate chambers?
- What is the maintenance interval and cleaning history in comparable installed sites?

Also, it's important not to over-rely on colour as a proxy for resistance. Light-coloured pavers reveal algae growth more clearly, but they aren't necessarily more susceptible unless poorly sealed, too smooth, or incorrectly jointed.


## Global Relevance

Algae-resistance is not just a tropical concern. Even temperate cities like London, Vancouver, or Seoul experience rapid algal colonization in shaded, poorly ventilated surfaces; particularly in urban canyons and near stormwater catchments. Green infrastructure strategies, which promote permeable and shaded landscapes, often unintentionally create micro-environments conducive to microbial growth. Without algae-resistant design, permeable paving and green-blue corridors can undermine their own safety objectives.

Countries such as Japan, Singapore, and Malaysia are advancing algae-resistant surface standards under local frameworks that integrate safety, maintenance, and climate performance.


## Final reflection

*"You can pressure-wash a sidewalk every week or you can specify a smarter surface once."*

This is not about luxury or premium product tiers. It's about applying what we know about biology, chemistry, and climate to design a surface that performs better, lasts longer, and protects both people and the environment. Because when a paving surface becomes high-maintenance, it's usually not the algae that failed—it's the specifier.

## 2.4 Modularity and End-of-Life Efficiency

*Why paving design should include its own exit strategy.*

Concrete is often treated as permanence—poured to last, shaped to endure. But in real-world conditions, every paved surface eventually reaches a point of intervention. A slab cracks. A section sinks. A service line needs to be rerouted. In traditional slab-based paving, this leads to one outcome: demolition. There's no graceful degradation—only jackhammers, dust, and removal.

The result is often significant amount of wate. Globally, construction and demolition debris accounts for nearly 40% of total solid waste, and non-structural elements like pavements form a large share of that footprint. When slabs fail, they rarely fail in isolation—but they're removed as if they did fail in isolation. However, the entire areas must be shut down, cut, lifted, and landfilled, only to be rebuilt using the same methods that led to failure.

Modular paving systems challenge that cycle. By design, they break the binary model of "intact or removed." Instead, they offer reversibility, selective repair, and functional reuse—not just during routine maintenance, but at the end of a project's life. This isn't just convenience. It's a deliberate engineering strategy that aligns with sustainability goals, operational continuity, and circular construction principles.

*"Modularity isn't just about flexible layout; it's about resilient performance over time and graceful disassembly when time runs out."*

### Repair Without Demolition

In a modular paving system, failure doesn't mean destruction. A dislodged or cracked paver can be lifted, inspected, and replaced within minutes—often using only hand tools.

There's no dust, no concrete slurry, and no shutdown of adjacent surfaces. Even sunken sections caused by subgrade settlement can be re-levelled and reinstated without disturbing utilities or uprooting adjacent zones.

This becomes critical in high-traffic or continuous-use settings such as industrial yards, airport service zones, bus terminals, or hospital loading docks; where full-slab replacement would mean extended downtime, rerouted circulation, and high logistical costs. Modularity enables micro-interventions: targeted, fast, low-disruption fixes that restore function without collateral damage.

**Operationally, this translates to**:
- Less disruption
- Lower maintenance budgets
- Greater uptime
- Less waste generated per repair cycle


## End-of-Life Efficiency: Reuse, Recycle, Redeploy

Perhaps the most overlooked advantage of modular paving is what happens when a site is decommissioned, reprogrammed, or redeveloped. In a conventional system, surfaces are scraped and discarded. But in modular paving, the materials themselves retain value, and can be removed, reassigned, or reincorporated.

- **Lift and relocate**: Pavers from a temporary installation or construction staging area can be reused in secondary applications or relocated to new projects.
- **Crush and reprocess**: Clean, uncoated pavers can be crushed and used as *recycled aggregates* in new concrete, road base, or as fill material—often exceeding 80% recovery efficiency.
- **Return to mix design**: In circular manufacturing setups, manufacturers may take back old units to integrate into fresh mixes, particularly when the product is mono-material and uncontaminated.

The result is a surface that behaves more like *a material asset* than a fixed installation. This idea of pavement as a *material bank* has gained traction in circular economy movements, with entities like the Ellen MacArthur Foundation calling for built assets to retain future use potential.

> *"Don't think of a paving system as something installed once.*
> *Think of it as a modular inventory stored in plain sight."*

## Aligning with Circular Economy Principles

Modular paving aligns tightly with the core tenets of circular construction, as defined by international frameworks such as:

- **The Ellen MacArthur Foundation**
- **ISO 20887:2020 (Design for Disassembly and Adaptability)**
- **EU Circular Construction Strategy**
- **UN Sustainable Development Goal 12: Responsible Consumption and Production**

These frameworks emphasize three pillars:
1. Design out waste and pollution
2. Keep materials in use
3. Regenerate natural systems

Modular systems check all three. They eliminate the need for full demolition. They preserve the reusability of discrete components. And they reduce the demand for virgin material production, which is among the most energy- and carbon-intensive processes in construction.

Several municipalities are now exploring *"digital material passports"* for hardscape assets—logging the batch data, sourcing, and potential reuse path of paving materials. Some even link this to green building credits, performance-based procurement, or carbon accounting.

## Design and Specification for Modularity

To fully leverage these advantages, modularity must be considered early in the design process, not retrofitted in. That means:

- Avoiding adhesive or mortar-bonded installations, which prevent disassembly
- Using edge restraints and bedding materials that maintain unit integrity under both installation and removal
- Choosing uniform sizes and thicknesses that allow cross-site compatibility and reduce reprocessing costs
- Specifying wear-tolerant textures and colours that age well, so that reused units aren't visually rejected

And just as important: documenting paving layouts, unit counts, and material sources. This allows future teams to understand what they're working with and plan future reuse intelligently.

***"A surface that can be uninstalled***
***is a surface that respects the future."***

Modular paving doesn't just deliver technical resilience. It delivers design intelligence. It prepares surfaces not just to perform but to evolve. Because long after a surface has served its first function, it may still serve another. All it takes is an engineer who planned for that from the start.


## 2.5 Material Efficiency and Embedded Carbon

*Why what goes into your paver matters as much as what it does on the surface.*

The sustainability conversation around paving tends to focus on what the surface *does*; how it manages water, handles heat, or reduces maintenance.

But the question that's asked far less often is: what did it take to make it? That's where the lens shifts from performance to embodied carbon; and suddenly, every kilogram of raw material, every batch of cement, every truck movement to the plant becomes part of the environmental story.

Embodied carbon refers to the total greenhouse gas emissions associated with a product's life cycle up to the point it's delivered to site. For pavers, this includes:

- Raw material extraction (limestone, clay, aggregates)
- Processing and clinker production
- Transport to manufacturing facilities
- Cement grinding, batching, and curing
- Packaging and site delivery

While a single interlocking paver may seem inconsequential in size, the sheer quantity installed across cities, campuses, highways, and airports makes their collective impact significant. In urban development, hardscapes often cover 20–50% of the total site area, and unlike vertical elements, they're rarely insulated, rarely lightweighted, and almost never modularized for embodied carbon analysis. But they should be.

According to the Global Cement and Concrete Association (GCCA), cement production alone accounts for roughly 8% of global $CO_2$ emissions, with OPC (Ordinary Portland Cement) responsible for ~850–900 kg $CO_2$ per tonne. This makes concrete mix design the most effective and immediate lever for reducing the carbon intensity of interlocking paving.

## Rethinking the Mix: Smarter Inputs, Cleaner Outputs

Lowering carbon in paving products begins at the material level, long before sustainability is mentioned in a marketing brochure. It involves rebalancing the mix to reduce clinker content, replace virgin materials, and optimize hydration chemistry.

## 1. Cement Substitutes (SCMs)

One of the most proven strategies is replacing OPC with Supplementary Cementitious Materials (SCMs)—industrial by-products that offer pozzolanic or latent hydraulic reactivity.

- **Fly Ash (FA):** A residue from coal combustion, rich in silica and alumina. Enhances long-term strength and workability, but availability is declining due to coal phase-outs in OECD nations.
- **Ground Granulated Blast Furnace Slag (GGBS):** A by-product of steel production. Offers high durability, sulphate resistance, and can cut embodied carbon by up to 50% when used at 50–70% replacement rates.
- **Limestone Calcined Clay Cement (LC3):** A newer class of binder using abundant, low-carbon materials. It lowers clinker factor while maintaining strength and early-age performance. LC3 can reduce $CO_2$ intensity by 30–40% relative to OPC mixes.

Each of these options reduces cement demand and *lowers peak hydration heat,* enabling energy savings during curing and extending service life due to reduced cracking and shrinkage.

## 2. Recycled Aggregates

Aggregates typically make up 60–80% of a paver's mass. Swapping virgin quarried stone for recycled sources significantly reduces environmental extraction costs.

- Crushed concrete and masonry waste can replace coarse aggregates, especially in base layers.
- Fine aggregates can be partially replaced with copper slag, bottom ash, or well-graded quarry dust; provided chemical stability and particle geometry are verified.

However, not all recycled aggregates are equal. Their performance depends on:

- Absorption rate and moisture content (affects water/cement ratio)
- Organic impurities or embedded steel
- Fines content, which may interfere with cement hydration
- Variability of particle size distribution (PSD)

For high-performance units, it's essential to pre-treat or screen recycled aggregates to meet strength and durability thresholds—typically aligned with ASTM C33 or EN 12620.

**3. Water-Reducing Admixtures and Strength Accelerators**
Chemical admixtures—particularly superplasticizers and early strength enhancers—allow for:
- **Lower water-cement ratios** without losing workability
- **Faster strength development**, enabling low-temperature curing
- **Shorter cycle times**, improving energy efficiency in automated production

These admixtures also allow for lower binder content per unit strength—another hidden but powerful carbon lever in the mix design toolbox.

## Environmental Gains Without Performance Loss
Critically, none of these improvements require a performance trade-off. When properly engineered, low-carbon pavers still meet or exceed all conventional benchmarks for:
- **Compressive strength** (e.g., >45 MPa for vehicular loads)
- **Abrasion resistance** (per EN 1338 or BS 6717)
- **Skid resistance**
- **Dimensional accuracy** and surface aesthetics

In fact, SCMs like GGBS and LC3 often improve sulphate resistance, chloride penetration thresholds, and long-term durability—factors that lower maintenance and replacement cycles, reducing *lifecycle emissions* in addition to up-front carbon.

*"Low-carbon doesn't mean low-performance.*
*If anything, it means the mix was engineered and not just poured."*

## Verifying the Claims: Data Over Declarations

Green marketing is easy. Substantiating those claims is harder, and non-negotiable in serious projects. To evaluate embodied carbon reductions in pavers, specifiers and regulators should demand:

- **EPDs (Environmental Product Declarations):** ISO 14025-compliant documents that quantify emissions using LCA methods (e.g., EN 15804 or ISO 14067). Prefer third-party verified.
- **Green labels** like:
    - MyHIJAU (Malaysia)
    - Declare (Living Future Institute)
    - GreenRE or GBI-certified concrete mixes
- **Mix design documentation** and compressive strength curves—particularly for government or large infrastructure contracts
- **PAS 2050 or ISO 21930-compliant lifecycle datasets** in public tenders

*"Without data, it's just marketing. With data, it's engineering."*

Some developers and governments now go further by setting carbon ceilings per $m^2$ of installed surface or embedding embodied carbon reporting clauses into tender documents. Many developed countries around the world are actively trialling carbon performance benchmarks for infrastructure elements, including non-structural components like pavers, kerbs, and edging.

Concrete is carbon-heavy—but not by design. It's carbon-heavy by convention. And as this section has shown, convention is increasingly being rewritten by engineers, not just environmentalists. When we rethink what goes into a paver—when we challenge every assumption about mix design, transport, and assembly—we begin to shift paving from being a carbon sink to being a low-impact, high-functioning surface system that delivers more with less.

## 2.6 Chapter Reflection

*What Makes Paving Truly Sustainable?*

Eco-friendly paving isn't defined by a checklist of green features; it is defined by a cohesive system response to real-world demands. It's not enough for a surface to absorb water or reflect heat. It must do so consistently, efficiently, and in context; while also using fewer resources, producing less waste, and creating safer, longer-lasting environments.

From how a paver is mixed, cured, and transported, to how it handles algae, dissipates heat, manages runoff, and eventually returns to the material cycle—sustainability lives in the details of its design and the integrity of its execution. The smartest systems are those that understand they are not isolated components, but part of a broader environmental, operational, and human landscape.

What we've explored in this chapter isn't just a catalogue of innovations—*it's a shift in mindset*. It's about designing paving systems not for installation, but for *performance over time*. It's about choosing materials not for appearance, but for *behaviour and longevity*. And it's about seeing every paver not just as a product, but as a decision with environmental, social, and financial consequences—one that echoes long after the surface is laid.

Because sustainable paving doesn't start with the first layer of stone. It starts with asking:

**"What kind of future are we building beneath our feet?"**

# Chapter 3: Smart Design for Sustainable Function

*How Layouts, Layers, and Logic Drive Performance*

Specifying the right material is only part of the equation. Too often, sustainability is treated like a box to tick at the product level, *high SRI? Check. Made with recycled content? Check.* But a truly sustainable paving solution isn't just made. It's engineered into place.

Eco-friendly pavers only perform as intended when they're part of a system; a system that considers how water moves, how load is transferred, how heat is managed, and how the surface responds to time, traffic, and environment. You're not just designing for day one. You're designing for ten years from now, when joints begin to shift, subgrades settle, or roots expand beneath the surface.

A permeable or low-impact paving system that looks good on paper can still fail in practice; often not because of the material itself, but because of what lies beneath it, or what was overlooked in its detailing. The slope may be off by half a percent. The joint aggregate might migrate during the first heavy rain. The subbase may be compacted too tightly to let water through. These aren't minor oversights; they're systemic weak points.

This chapter looks at the mechanics behind sustainable function; not from the perspective of materials alone, but as a whole system. How subbase design, slope, edge detailing, and layout patterning can either unlock the full potential of eco-friendly pavers or quietly set the stage for their failure.

Because in the real world, pavers don't fail in isolation. They fail with help.

## 3.1 Subgrade and Subbase Engineering

*Why everything above ground depends on what's beneath it.*

If the foundation is wrong, nothing else matters. That's not an exaggeration, it's a universal truth in pavement engineering. You can specify the best paver on the market, one with the highest compressive strength and the most impressive sustainability credentials, but if it sits on a poorly built subbase, it will fail. Quietly at first—then catastrophically.

Most failures in eco-friendly paving systems, particularly permeable ones, don't start with the surface. They begin below it; in the layers we often treat as "infrastructure" rather than "design." Cracking, ponding, heaving, algae accumulation, and sediment clogging often trace back to subgrade decisions that were rushed, underspecified, or misaligned with the surface system chosen.

A well-constructed base must do three things: hold water without softening, carry load without deformation, and breathe enough to stay permeable over time. That's not easy; especially in sites where the soil is variable, traffic is unpredictable, and water behaves like a force, not a feature.

In a typical cross-section, the sequence starts with a geotextile layer; an invisible but essential filter that prevents fine soils from migrating upward and clogging the system. Above that sits the open-graded subbase, often made of 300 – 700 mm crushed stone with high void content. This layer acts like a temporary reservoir, holding water during storm events and slowly allowing it to infiltrate. It also serves as the structural load carrier, spreading vehicle pressure evenly across the underlying soil.

On top of that is a base layer, composed of finer, angular aggregates that provide interlock and stability—flattening the pressure profile while preserving permeability. Finally comes the bedding layer: a slim, 20 – 50 mm zone where the pavers rest, levelled with precision and compacted just enough to allow for slight movement without displacement.

Every layer plays a role, and every shortcut adds risk. The integrity of a sustainable surface depends less on what it looks like, and more on what it's sitting on. Because in paving, beauty without bearing is just a performance waiting to collapse.

*"A permeable paver is only as permeable as the layer beneath it. Designing for infiltration isn't just about what drains; it is also about what holds, filters, and supports."*

## 3.2 Slope and Drainage Planning
*Why even permeable surfaces need a direction to flow.*

One of the most common misconceptions in sustainable paving is the idea that permeable equals flat; as if the ability to absorb water somehow eliminates the need for slope. It doesn't. In fact, that assumption is often where ponding, algae growth, and long-term structural problems begin.

Even in high-performing permeable systems, water needs a path. That means designing with positive surface slopes, typically between 1–2%, to guide excess water toward designated collection or infiltration points such as vegetated swales, bio-retention basins, trench drains, or hybrid systems that combine permeability with overflow control.

It's especially critical in mixed systems, where not all water is expected to infiltrate on site. In dense urban zones or high-traffic areas, where infiltration rates vary due to soil conditions or installation constraints, the paving system must serve as both filter and conduit. Water that doesn't go down must be directed somewhere else—intentionally.

In regions with intense or seasonal rainfall, slope becomes even more important. Flat permeable surfaces in such contexts can saturate quickly, turning joints into stagnant troughs rather than infiltration points. The solution lies in designing for dual function: letting water infiltrate when it can and directing it away safely when it can't.

That's where vegetated buffers, bioswales, and edge drainage channels come in—not just as aesthetic soft scaping, but as integral parts of the surface water strategy. These green infrastructure elements complement the paving by:

- Slowing and filtering overflow water
- Providing space for natural pollutant breakdown
- Reducing erosion at discharge points

*"Slope isn't a backup plan. It's the first layer of water strategy— especially when you expect the system to be permeable."*

When slope and drainage are designed with purpose, paving becomes part of the site's ***hydrological choreography***, not a surface reacting passively to rain. It's the difference between a surface that manages water and one that suffers from it.

## 3.3 Joint Width and Pattern Choices

*Why the spaces between pavers are where most of the engineering actually happens.*

In any paved surface, it's easy to focus on the units themselves; the finish, the texture, the size. But between those units lies the true technical battlefield: the joints. They might look like simple gaps, but they are the interface where water infiltrates, loads transfer, and the system either holds or slowly unravels.

The design of joints isn't just aesthetic. It has direct consequences for permeability, surface stability, long-term durability, and even maintenance costs. And the right choice depends entirely on what the surface is expected to do.

**Wider joints**, typically between 5 to 10 millimetres, are used when permeability is a primary goal. These joints allow rainwater to pass into the sublayers, helping with stormwater management. But width introduces a vulnerability: without the right joint infill, the system is prone to aggregate washout, weed growth, or structural loss during high-flow rain events. Angular, clean, crushed stone often in the 2–5 mm range is commonly used for its interlock and drainage properties. In exposed or high-maintenance areas, polymeric binders may be added to help stabilize the joint material without completely sealing it.

**Tighter joints**, in the 2 to 4 millimetre range, limit permeability but dramatically increase surface lock-up and rolling stability. These are ideal for applications like cycle lanes, factory floors, loading bays, or retail walkways—anywhere consistent alignment and surface comfort are critical. In these cases, joints act less like infiltration paths and more like mechanical couplings, distributing stress across the paver field to prevent localized displacement.

But joint width alone isn't the full story. The pattern in which the pavers are arranged plays a central role in how those joints perform over time. A herringbone layout, for example, doesn't just look traditional; it is one of the most mechanically resilient geometries for interlocking systems. The angular pattern distributes horizontal loads more evenly, especially in areas with turning wheels or dynamic pedestrian movement. That's why it's so often used in car parks, driveways, and warehouse apron zones.

In contrast, running bond or stacked bond patterns, while faster to install and visually minimal, offer less resistance to lateral force unless paired with effective edge restraints and low-tolerance paver dimensions. These patterns are better suited to low-traffic areas or pedestrian-only zones where aesthetics take priority over movement stress.

In stormwater-driven designs, joint and pattern decisions directly influence **void space percentage**, which is the total open area available for water to infiltrate. By adjusting joint width, selecting higher-void aggregates, or modifying pattern density, designers can fine-tune the system's **hydrological responsiveness** without changing the paver unit itself.

*"Designing a sustainable surface isn't just about choosing the right block—it's about engineering the space between them."*

And when joint detailing is neglected, the failure modes are predictable. Joints clog, surfaces shift, edges deform, permeability drops, and what was once a low-impact system becomes just another high-maintenance surface.

In paving, millimetres matter; especially when they're the only path water, force, and time must move through.

## 3.4 Edge Restraint: Holding the Line

Durability rarely unravels from the centre. More often, it fails at the margins; quietly, subtly, and then all at once. The edges of a paving system are where forces accumulate, stresses dissipate, and systems are either contained or allowed to drift apart. And yet, in many designs, edge restraint is overlooked; just relegated to a few lines in a drawing or left to on-site improvisation.

But make no mistake: the edge is a structural boundary, not a decorative one.

When a paver system carries load—whether from footfall, tyres, or trolleys; it transfers some of that energy laterally. Without sufficient confinement, the pattern begins to spread. Joints widen. Bedding shifts. And eventually, the field loosens. What starts as a fractional movement at the periphery becomes a systemic compromise.

In pedestrian paths, the signs are subtle: a faint curve at the boundary, a few pavers tilting outward. In high-load zones like turning areas in industrial yards or entry points for logistics vehicles, the results are more pronounced. Edges fracture. Corners chip. Interlock is lost. The surface appears misaligned not because the pavers moved independently, but because the edge failed to keep them together.

**"It's never just a broken paver. It's a missing boundary."**

Good edge restraint is not just about holding shape; it is about absorbing energy, distributing pressure, and maintaining order over time.

In lighter-duty applications, a well-formed concrete haunch, installed with sufficient embedment and proper drainage planning can serve well. But in zones with repeated vehicular stress, dynamic turning, or high moisture exposure, that simplicity falls short. Here, designers turn to mechanical edge systems: steel angles, precast channels, or pinned curb beams designed to work as part of the structure, not just next to it.

The material matters. But more critically, the continuity of the restraint system matters. Gaps at junctions, inconsistencies at grade transitions, or ad-hoc fixes during late-stage installation often open up pressure points. And like any pressurized system, it fails at its weakest link.

Edge performance also depends on what happens after installation. Uplift from root growth, undercutting from poor drainage, or erosion from stormwater runoff slowly degrade integrity. And yet in maintenance protocols, the edge is rarely checked, until the surface has already begun to deform.

A resilient paver system anticipates this. It designs its boundaries not just for appearance, but for mechanical function. It doesn't hope the edge will hold—it builds the edge to hold. Because no matter how strong the mix, how tight the joint, or how elegant the pattern; without restraint, everything moves.

## 3.5 Multi-Functionality: Beyond Just a Surface

*Why paving isn't just something you step on—it's something that works, thinks, and adapts.*

Modern paving systems are undergoing a quiet evolution (maybe even a revolution). Once considered passive infrastructure, just meant to carry weight, stay intact, and blend into the background; they are now being reimagined as active components in the built environment. They don't just support movement. They influence climate. They manage water. They shape public behaviour. In some cases, they even gather and respond to data.

This shift is being driven by necessity. As urban areas densify, space is at a premium. Designers and engineers can no longer afford to let surfaces do just one thing. Instead, we're seeing the rise of multi-functional paving systems that operate across technical, environmental, and even social dimensions.

At the material level, this begins with thermal management. Pavers with high Solar Reflectance Index (SRI) values contribute to localized cooling by reflecting solar energy, reducing the ambient temperature in plazas, sidewalks, and schoolyards. When paired with vegetated buffers, light-coloured walls, or canopy trees, this microclimate design can significantly reduce radiant heat loads—cutting cooling costs in nearby buildings and creating more liveable outdoor spaces.

But multi-functionality doesn't end with heat. Many eco-friendly systems now filter stormwater, not just direct it. When combined with proper subbase layering and geotextile separation, permeable paving can function as a biomechanical filtration system; where it removes sediment, heavy metals, hydrocarbons, and nutrients before runoff reaches aquifers or drainage channels. This move paving out of the "drainage problem" category and into the stormwater solution space.

And then there's the frontier: interactive paving systems. Smart cities are experimenting with embedded sensors that allow pavers to collect real-time data; on surface temperature, moisture content, traffic volume, or even dynamic load distribution. This data can be used to:

- Inform predictive maintenance
- Trigger alerts for clogging or heat stress
- Optimize pedestrian routing or emergency access
- Feed into building management systems or city-wide infrastructure dashboards

This kind of paving system doesn't just exist in the urban environment—it communicates with it.

Designers are also recognizing the behavioural and spatial role of pavers. Through variation in colour, texture, or layout, surfaces can signal intended use; inviting people into certain zones while guiding them away from others. A different pattern might cue a bike lane. A tactile strip may indicate a crossing zone. Slight changes in paver size or tone can break up long expanses and reduce the psychological fatigue of hard surfaces. In public space design, these are not minor details— they are non-verbal instructions embedded in the landscape.

***"Paving used to be about strength and durability. Now, it's about interaction between surface, climate, water, and people."***

This multi-functional approach reflects a broader shift: from designing isolated surfaces to engineering responsive environments. It demands interdisciplinary thinking, where civil engineers, landscape architects, sustainability consultants, and data technologists all have a stake in what happens underfoot.

The future of paving isn't flat. It's layered, responsive, and intelligent; working not just for vehicles or feet, but for the systems that cities rely on to stay liveable, efficient, and resilient.

## 3.6 Contextual Design: Not One Size Fits All

*Why smart paving isn't a product you specify once—it's a solution you adapt every time.*

There's a quiet danger in modern sustainability discourse: the assumption that good design can be standardized, copied, and deployed at scale without modification. But paving is not a universal plug-and-play solution; it is a localized intervention, and context is not a constraint — it is the foundation.

Climate, soil, maintenance practices, available materials, traffic load, and even the expectations of people walking across the surface— all of these vary from place to place. And when they do, they shift what "performance" means.

A paving system that excels in the temperate, moderate-rainfall environments of Northern Europe may crumble—literally and figuratively, under the punishing humidity of a Southeast Asian city. In a monsoon zone like Kuala Lumpur or Jakarta, water management and algae resistance are non-negotiable. Slope, drainage outlets, joint selection, and subbase void design become priority features. In these places, a high SRI rating may be a nice-to-have, but a poor edge restraint or an overly smooth surface can turn a courtyard into a slip hazard in a single storm.

Meanwhile, in arid, desert environments, the concerns flip. Rain is rare, but thermal stress is constant. Pavers must be UV-stable, resistant to brittleness, and capable of handling daily swings of 20–30°C. Dust, not algae, is the dominant contaminant. And joint stabilization becomes a priority; not for permeability, but to prevent dust accumulation and surface scour from wind abrasion.

Even soil behaviour plays a role. Clay-heavy soils in coastal regions require flexible systems that can tolerate slow heave and settlement. Sandy or loosely compacted soils demand broader load distribution and careful subbase compaction.

Permeable pavers designed for infiltration must consider local percolation rates, groundwater tables, and soil saturation thresholds, or they risk turning permeable systems into waterlogged liabilities.

And it's not just the physical world that varies. Maintenance culture is context too. In cities where municipal maintenance is reactive or infrequent, designing a surface that depends on regular joint cleaning is a recipe for failure. In contrast, a hospital, school, or corporate campus with a dedicated grounds team may be able to support a more delicate system, if the long-term cost savings are clear.

*"The best paving system in the world will fail in the wrong context; while the most basic one can thrive when perfectly matched to its environment."*

Even aesthetics is contextual. In heritage districts, modern concrete textures might clash with cobbled or sandstone surroundings. In cities with strong vernacular design language, like Kyoto, Jaipur, or Lisbon, the visual language of the pavers matters as much as their hydrological performance. People notice when surfaces feel out of place, even if they don't know why. Paving, in this sense, becomes a quiet part of cultural continuity.

This means good design isn't just about specs; it's about interpretation and negotiation:

- What's the priority in this climate?
- Who's maintaining it?
- What does failure look like here?
- What does success feel like to the people who will use this space daily?

Designing paving systems for sustainability means moving beyond best practices and into site intelligence. It means looking at paving not as a fixed product, but as a flexible framework; one that adapts to climate, community, budget, and behaviour.

The smartest paving solutions are never about repeating what worked last time. They're about learning from where you are, and building accordingly.

## 3.7 Chapter Reflection

In the end, sustainable paving isn't about isolated materials or eye-catching certifications; it's about designing systems that understand where they are, anticipate what they'll face, and endure without compromise. The smartest surfaces are not simply strong or porous or bright—they are those that have been thoughtfully integrated into the climate, culture, and cadence of their surroundings. They don't just sit in place. They perform in context. And when done right, they become more than infrastructure. They become infrastructure that works; for water, for people, for the long haul.

# Chapter 4: Greener Manufacturing, Smarter Materials

*How sustainability begins at the batching plant, not just the job site*

When people think of eco-friendly paving, their minds often go to performance; usually about how the surface handles water, reflects heat, or resists degradation. But behind every square meter of interlocking pavers lies a factory, and behind every factory, a sequence of material and energy decisions that shape the true environmental footprint of the product long before it's laid on site.

That's where the conversation about sustainability must expand. Because what happens inside the plant; how cement is sourced, how aggregates are processed, how curing is managed, how energy and water are consumed, sets the baseline for everything that follows. It's not enough to make a surface that performs sustainably. The manufacturing itself must be sustainable, verifiable, and adaptable to emerging environmental demands.

And yet, factory operations often remain in the shadows of green specification. Certification systems focus heavily on surface performance and site installation, while the upstream processes like batching tolerances, binder innovations, energy regimes, material traceability often goes unexamined. In a world that facing rapid decarbonization pressures and rising environmental disclosure standards, that's no longer tenable.

This chapter turns the spotlight inward; into the mixing chambers, curing racks, water loops, and conveyor belts of modern paver production. It examines not just how to build a better product, but how to run a cleaner, more efficient, and future-ready factory.

We'll look at binder innovations that reduce embodied carbon without compromising compressive strength. We'll explore the reuse of concrete waste and industrial by-products. We'll assess water loops, energy loads, emission points, and toxicity profiles. And we'll ground all of it in the reality of global manufacturing—where costs, consistency, and compliance intersect daily.

Because sustainability is not a feature you add at the end. It's a discipline you build in from the beginning. And for concrete pavers, that discipline starts not in the slab, but in the plant.

# 4.1 Cementitious Innovation
*Reducing Carbon Without Sacrificing Strength*

## A Material Under Pressure

Concrete has long held its place as the world's most used construction material, and cement has been its unchallenged binder. The material's reliability is undeniable; low cost, locally available, and capable of achieving high compressive strength in virtually any climate or application. For decades, the design conversation stopped there. As long as a concrete mix met the strength requirement and passed its durability test, its origin story was rarely interrogated.

That's changing, and rapidly too. With increasing scrutiny over carbon footprints, whole-life environmental performance, and material provenance, cement has become the industry's most visible vulnerability. Because while reinforcing steel, formwork systems, and even transport logistics have improved in efficiency, cement has remained largely unchanged in both chemistry and impact for over a century.

The problem is structural, not incidental. Cement's emissions aren't just due to energy inefficiencies—they're built into the reaction itself. When limestone ($CaCO_3$) is calcined to make clinker ($CaO$), it releases $CO_2$ as a chemical by-product. This decarbonation step accounts for over 60% of the emissions in OPC production. Even if kilns were electrified or fired with biomass tomorrow, a large chunk of the emissions would remain chemically locked into the process.

In the context of global net-zero ambitions; especially those aligned with the UN's Race to Zero, the European Green Deal, or the Global Cement and Concrete Association's 2050 roadmap—cement's environmental burden can no longer be treated as fixed. It must be treated as an engineering challenge, one where innovation is required not only at the lab scale but across industrial production lines.

That puts precast products like pavers at the centre of this evolution. Unlike structural concrete, pavers offer more control over curing, mix design, and batching conditions. They present a unique opportunity to test and scale cementitious alternatives that dramatically reduce embodied carbon, without waiting for the entire construction sector to catch up.

## Reducing Clinker, Not Performance

The first and most immediate lever is clinker substitution. By partially replacing OPC with supplementary cementitious materials (SCMs), manufacturers can significantly reduce emissions without changing production lines or compromising output quality.

**Ground Granulated Blast Furnace Slag (GGBS)** has emerged as a workhorse SCM. As a by-product of steel manufacturing, it carries a fraction of OPC's embodied carbon, while improving long-term strength, sulphate resistance, and durability. In well-designed mixes, GGBS can replace up to 70% of cement by weight—cutting carbon while extending service life.

**Fly Ash**, another industrial by-product, has long been valued for improving workability and long-term pozzolanic activity. But as the world transitions away from coal, high-quality fly ash is becoming less accessible; especially in Europe and parts of Southeast Asia.

**Limestone Calcined Clay Cement (LC$^3$)** represents a next-generation solution, particularly in regions with abundant kaolinite-rich clays. Developed by EPFL and increasingly adopted across India and Latin America, LC$^3$ reduces clinker content by up to 40% without sacrificing 28-day strength or requiring specialized curing.

These materials don't dilute concrete—they enhance it, when properly used. They reduce early hydration heat, refine pore structure, and often improve durability against sulphates, chlorides, and carbonation; all of which translate to longer intervals between maintenance or replacement.

## Beyond Substitution: Reinventing the Binder

*Why the next frontier in concrete isn't about what's removed—it's about what's reimagined.*

Replacing part of the cement content with industrial by-products is no longer a fringe idea—it is the new normal. Fly ash, GGBS, and calcined clays have earned their place in standards, specifications, and procurement checklists. But these strategies, while impactful, still operate within the traditional binder paradigm. They mitigate emissions, but they don't eliminate the source. The next generation of innovation demands more: a structural shift in how binding agents is conceived, formulated, and activated.

This is where alkali-activated materials (AAMs) and geopolymer systems step in; not as incremental improvements, but as conceptual departures. Unlike traditional hydraulic cement, which hardens through hydration, these systems rely on polymerization of aluminosilicates triggered by alkaline activators.

The result is a hardened binder with zero clinker, potentially zero limestone, and drastically lower embodied carbon.

Field data from Australia, the UAE, and India has shown that geopolymer pavers produced using high-calcium fly ash or GGBS activated with sodium silicate, can exceed 60MPa in compressive strength, offer improved resistance to sulphate and acid attack, and reduce GWP by 60–80% compared to OPC-based equivalents. Some even claim net-zero $CO_2$ when local industrial waste is used as a precursor.

But challenges remain. Chief among them:
- **Standardization:** There's no universal code for geopolymer binders. Efforts by RILEM and ASTM are still in draft form.
- **Chemical handling:** Activator safety (e.g., NaOH) poses logistical and occupational health risks.
- **Supply chain maturity:** Unlike cement, geopolymer production requires new procurement and QC pathways.
- **Durability dataset:** While lab results are promising, field performance data beyond 10 years remains sparse in high-load or extreme climates.

Despite these limitations, the potential is undeniable. In precast environments, where curing, quality control, and exposure are tightly managed, these systems offer a viable route to clinker-free construction.

More practically, even without going fully binder-free, some manufacturers are exploring ternary and quaternary blends, by combining OPC, GGBS, calcined clay, and limestone powder; and that reduces clinker by up to 50% while maintaining hydration synergy.

The advantage is that these blends:
- Fit within existing batching systems
- Rely on widely available materials
- Are compatible with established curing processes
- Can be tailored to regional material constraints and performance needs

*"The innovation isn't always in total replacement,*
*It is in how the chemistry is balanced, activated, and scaled."*

This is the evolution of cementitious thinking. It's not just about removing cement; it's about designing next-generation binders that perform, endure, and emit less from the first bag to the final placement.


## Strength, Yes. But Smarter

*Why durability is not the trade-off in low-carbon concrete—it's the proof of engineering.*

For years, the biggest barrier to adopting low-carbon cementitious systems wasn't technical, but rather psychological. The moment a designer or contractor hears "reduced cement content," the reflexive fear is that strength will suffer. That performance will lag. That failure will follow. But these fears reflect a legacy mindset; one that rooted in over-designed mixes, oversized safety margins, and decades of treating cement as a one-size-fits-all insurance policy.

Today, the relationship between strength and sustainability is no longer inverse. In fact, the more advanced the binder system, the more carefully optimized the mix, the more consistently performance is achieved—not just in the lab, but on the line and in the field.

## Replacing Mass with Precision

Traditional high-OPC mixes were designed for simplicity and margin. The assumption is that; add more cement than necessary to overcome variability in curing, aggregate quality, and workmanship. But that approach introduces its own sets of problems like shrinkage, thermal cracking, durability issues in sulphate or chloride-rich environments.

Modern SCM-enhanced mixes or ternary systems are the opposite. They are designed to work synergistically. Each component is selected for a role:

- Slag for long-term strength and sulphate resistance
- Fly ash for particle packing and pozzolanic reactivity
- Calcined clays for early-age strength and water retention
- Fine limestone for rheology and filler interaction

The result isn't a weaker paver. It's often a stronger, more durable one, with fewer defects and longer life between replacements. In multiple comparative trials (e.g., $LC^3$ vs OPC, 50% GGBS mixes vs OPC), 28-day compressive strength regularly meets C35–C50 classes, and flexural strength remains well within interlocking application norms.


## It's Not Just Strength—It's Resilience

Strength, of course, is only one metric. Durability tells the full story:
- Abrasion resistance (EN 1338: Annex D): SCM-rich pavers typically show lower wear loss, especially in slag or silica fume blends.
- Freeze-thaw cycles (ASTM C1262): Lower water absorption improves performance in temperate and continental climates.
- Water permeability: Denser pore structure with finer SCMs reduces ingress and increases resistance to freeze damage and de-icing salts.
- Surface hardness and skid resistance: Remain equal or superior with proper curing, pigment stabilization, and surface compaction.

## Field-Proven, Not Just Lab-Tested

Low-carbon mixes aren't experimental anymore. In projects from Denmark to Delhi, precast pavers made with up to 60% GGBS or 30% LC$^3$ have been in service for over a decade, with minimal signs of degradation under pedestrian and low-speed vehicular traffic. Performance tracking across airports, institutional campuses, and green-rated developments confirms:

- No significant difference in structural failure rate
- No increase in efflorescence or discoloration with proper curing
- Comparable or improved lifecycle maintenance profiles

*"Strength, in modern concrete, is no longer about excess—it's about calibration. Low-carbon performance is not the compromise. It's the evidence of better engineering."*

So the question is no longer, "Will this hold up?" It's "Why are we still relying on mixes that haven't evolved since the 1960s?"


## The Forgotten Variable: Curing Energy

*Why sustainability doesn't stop at the mix—it continues through the chamber.*

You can design the perfect mix on paper; a balanced SCM ratios, optimized aggregate grading, a water-cement ratio tuned to the decimal. But if the curing process is inefficient, uncontrolled, or misaligned with the binder chemistry, all of that engineering falls short. Worse, it adds unseen emissions to the final product.

Curing is where concrete becomes concrete. And in the case of precast pavers, it's often the single most energy-intensive step in the production chain. In plants relying on rapid turnaround cycles, steam-curing systems push heat into chambers for 8 to 18 hours, sometimes longer.

The fuel behind that heat, such as diesel, LPG, natural gas, or high-draw electricity; isn't just a line item on the utility bill. It's part of the product's carbon footprint.

Studies across Europe and Southeast Asia show that curing energy alone can account for 15–25% of the total embodied carbon of a concrete paver; especially where elevated temperature curing is used to meet 24-hour strength targets. That's why curing should never be treated as just a technicality. It's an opportunity.

Innovative manufacturers are now investing in closed-loop thermal systems, where exhaust heat from boilers is recovered to preheat incoming water or air, reducing primary energy demand. Others are trialling carbon-injection curing, where captured $CO_2$ is introduced into the chamber and reacts with calcium hydroxide in the mix—locking carbon into the matrix as stable calcium carbonate. This process doesn't just reduce emissions. It turns the curing cycle into a sequestration step.

In warmer climates, where ambient curing is viable, SCM-rich mixes (particularly GGBS or $LC^3$ blends) are extending curing times slightly to enable low-energy hydration curves; thus, achieving target strength with minimal or no external heat.

Even the design of the mould matters. Moisture-retentive or insulated formwork helps maintain surface hydration and internal RH without misting or thermal input. That means better pigment stability, fewer early-age cracks, and lower energy intensity.

But perhaps the biggest shift required is philosophical: to treat curing not as a background process, but as a critical component of sustainable mix performance. If we're serious about reducing embodied carbon, then what happens in the curing chamber must be as tightly engineered as the binder itself.

*"The smartest concrete isn't just what's in the mix. It's what happens after it's poured and how little it takes to make it strong."*

## From Mix Design to Market Trust

*Why data is the foundation of credibility in low-carbon innovation.*

The promise of low-carbon concrete, especially in paver manufacturing, is no longer theoretical. The mix designs exist. The chemistry works. The strength benchmarks are being met every day in factories around the world. And yet, despite all that, scepticism remains.

Why? Because in an industry still dominated by performance guarantees, contractual liabilities, and conservative design codes, narratives don't build trust—numbers do.

This is where the next layer of innovation must happen, not in the binder, but in the validation and communication of performance. In a market increasingly shaped by embodied carbon targets, circular procurement clauses, and lifecycle declarations, manufacturers must be able to move seamlessly from "here's what we made" to "here's what it cost the environment to make it."

That's where Environmental Product Declarations (EPDs) steps in; not as a marketing checkbox, but as a technical passport. A well-prepared, third-party verified EPD tells a specifier how much $CO_2$ was emitted per square meter of product, how materials were sourced, what energy was used in curing, and what impact those decisions had across a cradle-to-gate (or full LCA) analysis.

It tells the truth about what was engineered.

It also opens access to projects where carbon tracking is now mandatory such as in public works tenders in Sweden, digital product passports in the Netherlands, Green Mark Platinum projects in Singapore, PAS 2080-aligned infrastructure in the UK, or MyHIJAU procurement in Malaysia. In these markets, proof of carbon competency is no longer a differentiator. It's a requirement.

But market trust also extends beyond data sheets. It lives in repeatable performance. In field results that align with lab trials. In site installations that hold up after ten monsoons. In clients who returns; not because of the sales pitch, but because of the surface they're still walking on.

Because ultimately, the credibility of a low-carbon paver doesn't come from what it claims. It comes from how well it was made, and how clearly that can be shown.

*"The strength of the product will win the contract.*
*The transparency of the process will win the future."*


## Final Thought

*From mix to method, strength is no longer the only benchmark.*

We used to ask: "Will it hold up?" Now we must ask: "What did it cost to make it hold up?". Cementitious innovation isn't just about new materials, it is about shifting values. It asks us to reconsider the assumptions we've built entire systems upon, that more cement means more strength, that faster curing is always better, that performance is separate from process.

But performance *is* process. And sustainability is no longer a trade-off. It's an indicator of how deeply we understand the material, how well we control our operations, and how seriously we take our role in shaping the built environment's future.

The pavers we produce today aren't just products. They're reflections of chemistry, energy, intention, and accountability. And the manufacturers who treat every mix as a moment to improve; not just the surface, but the system, are the ones leading the industry forward.

*"Concrete will always be strong. Now, it must also be wise."*

## 4.2 Recycled Aggregates and Waste Inclusion

*Turning yesterday's concrete into tomorrow's resource.*

Concrete has always had an appetite for raw material. Every cubic meter demand hundreds of kilograms of sand, gravel, crushed stone, or industrial fines aggregates that must be extracted, transported, washed, and graded before they ever reach the mixer. On a global scale, this appetite adds up fast: the construction sector consumes an estimated 50 billion tonnes of aggregates each year, making it the most resource-intensive industry on Earth by mass.

Interlocking pavers might be small individually, but collectively they cover millions of square meters across roads, plazas, ports, and parks. Which means that even modest gains in aggregate efficiency, especially in high-volume precast production can yield enormous environmental benefits.

This is where recycled aggregates and industrial by-product inclusions come into focus. They are not just waste diversion strategies. They are essential pathways to a circular, resource-conscious model of concrete manufacturing—one that reduces extraction, minimizes landfill, and reimagines construction debris as feedstock, not burden.

But effective reuse is not about dumping rubble into a mixer. It requires intention, processing, testing, and specification clarity. It is a technical process, not a feel-good gesture.

## The Aggregate Opportunity

*Why reuse is no longer just an environmental ideal—it's a design imperative.*

The conversation around aggregates has always focused on availability, gradation, and strength. But as the pressure on natural resources intensifies, that conversation must evolve.

Today, aggregate sourcing is not just about what works technically; it's about what works ethically, economically, and ecologically.

Every interlocking paver contains more aggregate than binder. In fact, aggregates make up 70–80% of a paver's total mass. Which means if you're looking to cut the environmental cost of concrete, there's no bigger lever. And yet, in most manufacturing plants, the source of that aggregate still traces back to virgin quarries; sites where rock is blasted, crushed, washed, and trucked for miles before it ever sees a mixer.

This isn't sustainable. And it isn't necessary.

Across the globe, construction and demolition (C&D) waste has emerged as an untapped mineral reserve. Crushed concrete, discarded masonry, roof tiles, and even select ceramics, when correctly processed can function as viable aggregates. Same goes to industrial by-products like steel slag, quarry fines, bottom ash, and copper slag. These aren't just waste streams. They're resource streams, if we have the systems to convert them.

The real opportunity is not in chasing exotic materials or rare technologies. It's in building local circularity: taking what's already coming out of demolition sites, factories, and infrastructure retrofits— and closing the loop within the region's own paving ecosystem.

In a world where the UN forecasts global sand scarcity by 2050, and cities like Singapore, Amsterdam, and Tokyo are implementing material passports and reuse mandates, this isn't a future trend. It's the new baseline.

*"The smartest quarry may not be in the hills,*
*It may already be under your feet."*

## Performance Considerations: Quality In, Quality Out

*Why recycled aggregates demand more engineering—not more compromise.*

The success of recycled aggregates isn't found in blind substitution. It's found in the precision of the process. For all the sustainability potential they offer, these materials are not inherently equal to their virgin counterparts; not because they are weaker, but because they are more variable. And variability, in a tightly controlled manufacturing environment like interlocking paver production, must be understood, calibrated, and managed properly.

Concrete doesn't fail because a material was recycled. It fails because the material wasn't characterized.

The challenge with recycled aggregates is not technical incompatibility; it's the inconsistency. Crushed concrete from a demolition site might contain remnants of mortar, high porosity, or embedded steel. Industrial slags may offer excellent angularity and strength but carry risks of expansive reactions or heavy metal traces if improperly processed. Even within a single feedstock stream, the physical and chemical properties can fluctuate by batch or source.

That's why recycled aggregates must be treated not as *waste materials* given a second chance, but as engineered components requiring first-class quality control.

In high-performance paver production, this means:
- **Particle size distribution** must be predictable—not just on paper, but per load. Poor gradation alters packing density, water demand, and compaction response.
- **Porosity and water absorption** must be tested—especially for fines. These values directly affect mix water calculations and w/c ratio stability.

- **Contaminants** must be screened—not occasionally, but systematically. Even trace amounts of gypsum, asphalt, organics, or chloride-laced debris can undermine mix stability or long-term durability.
- **Mechanical integrity** must be validated, not assumed. Impact resistance (e.g., via Los Angeles abrasion or aggregate crushing value tests) confirms that recycled aggregates can handle the compaction and load expectations of interlocking systems.

And critically: moisture content must be known at the point of mixing. Unlike natural aggregates, recycled materials often arrive with highly variable surface and internal moisture, especially if stockpiled without cover. Pre-wetting or saturation protocols may be needed to avoid hydration shocks or batch-to-batch slump inconsistency.

Leading plants address this through:
- Preprocessing lines with magnetic separators, washing screens, and dual-stage crushers
- On-site material testing labs, capable of same-day grading and absorption assessments
- Digital QC systems that log source, processing, and performance data per load

These are not over-engineering exercises. They are the cost of building trust in a material class that still faces legacy stigma.

*"Recycled doesn't mean reduced quality. It means re-qualified, re-tested, and re-applied with intent."*

In the end, the goal isn't to fit recycled materials into conventional systems by hope or habit. It's to engineer systems that recognize their strengths, manage their variabilities, and deploy them precisely; because quality in, truly is quality out.

## Environmental and Economic Impact
*Why circularity isn't just greener—it's leaner.*

Every aggregate pulled from the earth carries a cost, and not just the one on the invoice. The real cost is ecological. Quarrying strips vegetation, scars landscape, disturbs groundwater, and adds truckloads of emissions through blasting, crushing, washing, and transport. For decades, these costs were treated as externalities. But in a carbon-constrained world, and an increasingly urbanized one, those impacts are now being brought back onto the balance sheet through environmental regulation, embodied carbon accounting, and the rising cost of natural resource extraction. Recycled aggregates shift this equation, not just environmentally, but economically.

From a sustainability standpoint, the benefits are substantial. Each tonne of crushed concrete or industrial slag that substitutes virgin stone represents:
- Fewer kilometres driven by diesel-powered dump trucks
- Lower water consumption in aggregate washing and grading
- Reduced pressure on depleting natural sand and gravel deposits
- A direct decrease in landfill dependency and tipping volume

In many lifecycle assessments (LCAs), recycled coarse aggregates show an embodied carbon reduction of 20–40% compared to virgin equivalents; depending on local transport distances, electricity mix, and processing intensity. These reductions don't require new technology. They just require intention and infrastructure.

But the economic advantages are equally compelling.

Where logistics are well-managed and supply is local, recycled aggregates can reduce input costs significantly; particularly in urban areas where virgin material prices are inflated by haulage or scarcity.

This is especially true in jurisdictions with:
- Green procurement incentives, where recycled content unlocks scoring bonuses or fee reductions
- Waste diversion targets, where construction firms are penalized for landfilling C&D debris
- Carbon-linked tenders, where low-GWP products are favoured in selection criteria

Some manufacturers have turned this into a differentiator by openly declaring recycled content as a percentage of total product mass on product sheets, EPDs, or digital passports. In the Netherlands, it's increasingly common for interlocking systems to carry "circularity ratings," indicating the proportion of reused material and reusability at end of life.

At a systems level, the circular use of aggregates also leads to:
- Reduced stockpiling of demolition debris
- Co-location synergies between crushing plants and precast yards
- Lower storage footprint and batching variability, when aggregate streams are stabilized

What these reveals is a powerful alignment: what reduces environmental harm also reduces operational friction. Waste becomes input. Input becomes optimized. And the boundaries between disposal, processing, and production blur into a more elegant model of material management.

*"The goal isn't just to reduce harm. It's to create a resource cycle so efficient, waste no longer has anywhere to go."*

## Applications and Limits
*Where recycled aggregates shine—and where engineering caution must still lead.*

Recycled aggregates are not a universal solution—but they are a powerful one. Like any material system, their success depends not just on composition, but on context. The key is not to force recycled content into every application, but to know exactly where it fits, why it works, and where it needs to be restrained. In interlocking paver production, that clarity matters. These units aren't poured in place—they are precision products. Surface texture, edge fidelity, abrasion resistance, and dimensional stability all play critical roles in field performance. Which means every aggregate input must contribute not only to environmental savings, but to structural and aesthetic consistency as well.

## Where Recycled Aggregates Excel
Recycled aggregates perform exceptionally well in:

- Base layers and bedding courses, where surface exposure is minimal, and strength demands are predictable
- Non-architectural or pigmented pavers, where minor colour variation is acceptable or masked
- Secondary or peripheral zones like pedestrian footpaths, garden edging, light-duty parking bays
- Projects seeking green certifications, where documented recycled content supports LEED, GreenRE, or GBI criteria
- High-volume, low-variability production runs, where recycled material supply can be stabilized and QC tightly integrated

In these settings, the performance of recycled aggregates can match, or even surpass, that of virgin equivalents; especially when slag-based aggregates or well-crushed concrete fines are used. Their angularity often improves interlock behaviour. Their embodied carbon savings add up fast. And their supply proximity reduces haulage emissions at scale.

## Where Caution is Still Warranted

But there are also zones where recycled aggregates require constraint, adaptation, or exclusion; not because they're inferior, but because the design demands are too narrow to tolerate variability.

Engineers should proceed carefully in:
- Architectural finishes or decorative units, where surface consistency, aggregate exposure, or tone matching is critical
- Freeze-thaw or de-icing environments, where retained moisture in porous recycled particles can trigger spalling
- High-friction zones, such as loading docks or turning radii, where impact resistance and abrasion loss must be tightly controlled
- Poorly regulated supply chains, where contamination risk (chlorides, sulphates, wood, gypsum, glass) remains high and difficult to verify
- Thin-section or low-clearance moulding, where edge retention and compaction sensitivity are paramount

In these contexts, the better strategy may be to deploy recycled aggregates below the surface, or to blend them at lower substitution rates; thus, preserving sustainability gains without compromising final product performance.

Ultimately, circularity in aggregate sourcing is not about absolute adoption. It's about engineered allocation. Knowing where the line sits and where recycled materials offer true functional value is the mark of a mature, adaptive production system.

*"The future of paving isn't made from one material.*
*It's made from a thousand smart decisions;*
*measured, placed, and layered with purpose."*

## Final Thought

*Concrete is not a singular material—it's a system of choices.*

Every paver we produce is a reflection of resource decisions: where our materials came from, how they were processed, and how much we respected the cycle they emerged from. Recycled aggregates don't weaken that story; they only deepen it further. They force us to be more precise, more intentional, and more engaged with the full lifecycle of what we build.

Because sustainability doesn't begin with the product. It begins with what we allow to become material in the first place. Waste isn't the enemy of quality, but neglecting it is. And the more we learn to engineer from what already exists, the less we'll need to extract from what remains.

## 4.3 Energy Use in Paver Manufacturing

*Why operational efficiency is the next frontier of sustainability.*

Sustainability is often framed in terms of materials; what's in the mix, where it came from, and how it's reused. But behind every eco-friendly paver lies a factory. And behind every factory is a complex network of machines, motors, conveyors, compressors, kilns, and curing systems—all of which run on energy. If we ignore this part of the equation, we risk calling a product sustainable simply because the ingredients are cleaner; even when the process that makes it continues to burn through kilowatt-hours at scale.

In interlocking paver manufacturing, energy isn't just a utility cost—it's a carbon vector. It determines not just operational efficiency, but the legitimacy of embodied carbon claims, the credibility of Environmental Product Declarations (EPDs), and a manufacturer's alignment with national and global decarbonization goals.

More importantly, it's one of the few levers where emission reductions and cost savings align perfectly, making energy management not just a sustainability imperative, but a strategic business opportunity.


## Where the Energy Goes

*Understanding the anatomy of energy consumption in paver production—and why it matters more than ever.*

Before any discussion on efficiency or decarbonization can begin, we must first understand the energy architecture of a typical interlocking concrete paver plant. Energy flows through these facilities in both obvious and hidden ways, powering not only production but the entire support ecosystem that surrounds it. Misunderstand where that energy goes, and you risk optimizing the wrong system—or worse, ignoring the biggest carbon culprits entirely.

Let's break it down; not from a utility bill perspective, but from the point of view of energy embodied in the product.


## 1. Batching and Mixing Operations

Every paver begins with dosing and mixing. This phase includes:

- Conveyor belts transporting aggregates to hoppers
- Precision weighing of cement, water, SCMs, pigments
- High-shear mixers, often pan-type or planetary, blending the mix uniformly
- Material transfer to moulding stations

While each step may seem incremental, collectively they constitute a major base load on the plant's electrical system. Motors are often oversized for reliability, running on direct online starters instead of energy-efficient VFDs. Mixers, in particular, can draw 15–30 kW per cycle depending on capacity, with peak surges during dry mix initiation.

Poor batching sequence planning can further inflate this footprint; such as restarting mixers between every cycle or maintaining conveyor idle running during breaks. In plants running two or three shifts per day, cumulative inefficiencies compound rapidly, contributing both to energy waste and thermal load inside the plant.

## 2. Curing Systems

This is typically the largest energy consumer in the entire process—and often the least optimized.

In high-throughput plants, pavers must be strong enough for demoulding within 6–24 hours. To achieve this, manufacturers rely on:

- Steam curing chambers powered by gas/diesel-fired boilers
- Electric curing cabinets with resistive heating
- Hydronic systems using hot water pipes beneath curing racks

These systems aim to elevate internal mix temperatures to 50–80°C, accelerating hydration and strength gain. But the energy demand is immense. A single 500 m² curing chamber running at 60°C for 8–10 hours can consume 250–500 kWh per cycle—or more if heat loss is unmanaged.

The problem isn't just the heat. It's how that heat is managed:
- Poor insulation = continuous energy bleed
- Uneven heat distribution = inconsistent strength and overcured zones
- Overly aggressive ramping = increased microcracking and shrinkage, leading to performance loss downstream

## 3. Auxiliary Systems

These are the energy consumers that hide in plain sight:
- Compressed air systems for mould release, cleaning, and pneumatic controls
- Hydraulic power units in press-mould systems
- Water pumps for aggregate washing or wet curing

- Dust extraction fans, lighting, and HVAC in office/control rooms
- Idle load draw from PLCs, control panels, or uninterruptible power supplies

While each system may consume small amounts of energy in isolation, collectively they form what energy auditors call the parasitic load; a persistent base load that often goes unnoticed and unmonitored. Worse, compressed air systems are notoriously inefficient: studies show up to 25–30% of air produced is lost to leaks, and improperly maintained compressors can draw double their rated power under load surges.

## 4. Why This Breakdown Matters

If you're calculating embodied carbon or even just trying to reduce your energy bill, knowing where the energy goes is the first step. More importantly, it guides where interventions will have the greatest return:

- Optimizing mixers and batch cycles? Valuable—but incremental
- Improving curing efficiency? Transformative
- Tightening up auxiliary systems and parasitic draw? Often the fastest ROI available

And in markets where carbon-linked procurement or green tax incentives are tied to product-level energy intensity, this breakdown isn't academic—it's contractual. You don't just need to save energy. You need to prove where, when, and how it was saved.

*"In the era of net-zero targets and carbon benchmarking, understanding your energy anatomy is no longer optional. It's the start of every meaningful sustainability conversation."*

## Measuring the Invisible: Carbon per Kilowatt

*Why a kilowatt-hour is no longer just a number—it's a declaration.*

Walk into any paver factory and ask how much electricity is used per cycle, and you'll get a number. Maybe it's in kilowatt-hours per tonne of concrete, maybe per shift. But ask how much carbon that electricity represents, and the answers begin to falter.

That's the strange truth of energy use in modern manufacturing: we track consumption meticulously but rarely translate that into the metric that now matters most—***embodied carbon***.

And yet, carbon is what policymakers are regulating, what certifiers are measuring, and what project developers are increasingly asking for. Whether it's under Malaysia's *Low Carbon Cities Framework*, the EU's *Construction Products Regulation*, or global tools like *EN 15804* or *ISO 14067*, the question isn't just: "How much energy did you use?" It's: "What did that energy mean for the atmosphere?"

That meaning changes by region. In Denmark, a kilowatt-hour of grid electricity might carry just 50 grams of $CO_2$. In Thailand or parts of Malaysia, that same kilowatt-hour could come with over 600 grams; that's about 12 times the footprint. The number doesn't come from the machine. It comes from how your grid is powered: coal, gas, hydro, solar, or an unsteady mix of them all.

This variability turns embodied carbon into a location-sensitive metric. Two paver factories which is identical in process, identical in product can produce drastically different footprints purely because of geography. A curing chamber in Penang running on coal-heavy grid power will emit far more than the same system in Berlin powered by renewables, even if energy use is identical.

So, what does this mean for producers?

It means that knowing your kilowatt-hours is no longer enough. You must know what those kilowatt-hours represent. Not just theoretically, but numerically. Verified. Declared. Auditable. Because in low-carbon procurement, the difference between a job won and a job lost may be a matter of grams per square meter.

And the tools to do this already exist. Environmental Product Declarations require it. PAS 2080 expects it. Government tenders increasingly demand it; not as an innovation, but as baseline disclosure.

This is where a deeper transformation begins: when energy tracking stops being a maintenance function and becomes an engineering language. When plant managers don't just report kilowatts, but can model how those kilowatts flow through the curing system, the mixer cycles, the auxiliary loads. And when that model isn't buried in an Excel sheet but embedded in the company's promise to the market.

Because in the age of carbon-conscious construction, every kilowatt tells a story. Not just about what was built but about how responsibly it was built.

*"If you can't measure it, you can't defend it.*
*And if you can't defend it, it won't qualify."*

## Operational Strategies for Energy Reduction
*Why reducing energy isn't about shortcuts—it's about system intelligence.*

Energy efficiency in paver manufacturing has long been mistaken for a checklist: install a few LED lights, insulate the curing chamber, clean the air compressor filters, and call it a day. But efficiency isn't a retrofit. It's a mindset—a way of reading the factory floor not just for output, but for waste disguised as habit.

Some of the most energy-intensive behaviours in a plant are not malicious—they're just inherited. Motors left humming through lunch breaks because "that's how it's always been." Mixers oversized for their current batch sizes, running at full draw for half-capacity loads. Curing cycles timed for the worst-case mix, not the actual hydration profile. These aren't faults in technology. They're blind spots in practice. True operational energy reduction doesn't begin with equipment upgrades. It begins with asking: where is the energy going, and does it need to go there?

Take batching, for example. When mix schedules are optimized—grouped by type, volume, and curing requirements, the energy demand flattens. Load surges decrease. Idle time drops. There's no new machine involved just better sequencing, clearer planning, and fewer dead cycles. A smart factory doesn't just automate mixing—it times it with intent.

Or consider curing, the thermal elephant in the room. Instead of applying uniform heat to every paver, regardless of chemistry, modern plants are using real-time feedback from hydration models. GGBS-rich mixes, for instance, may require longer but cooler curing curves. Steam is reduced. Energy spikes are avoided. And durability often improves as a side effect. It's not efficiency for its own sake—it is performance aligned with energy logic.

Even the humble conveyor belt is part of the conversation. Most are powered by direct-drive motors that run at fixed speeds. But do they need to? Variable frequency drives (VFDs), matched to real-time load sensing, adjust speed dynamically, reducing draw without stalling throughput. Over time, that difference accumulates; not just in kilowatts saved, but in lower wear, less vibration, and better control.

And then there are the systems no one thinks about: air compressors, water pumps, ventilation units. These background actors often run inefficiently simply because no one's watching. But in energy audits across Germany, Japan, and Malaysia, these "support" systems regularly emerge as 15–25% of total plant electricity use. When compressed air is leaking from 10 joints or when pumps run 24/7 to fill a tank used 6 hours a day—that's not just inefficiency. That's money slipping through a gap no one measured.

But here's what matters: none of these solutions require sacrificing throughput, quality, or reliability. Quite the opposite. Plants that integrate energy intelligence into operations tend to gain predictability, longer machine life, and more consistent product outcomes.

Because reducing energy isn't about trimming fat. It's about understanding the metabolic health of the entire production system. It's about tightening the loop; not just for carbon compliance, but for operational resilience in a market that rewards both.

*"Efficiency isn't about doing less.*
*It's about doing exactly what's needed and nothing more."*

## Renewable Integration: From Optional to Operational
*Why the factory roof is now a climate strategy—and a competitive one.*

For years, renewable energy in manufacturing was treated like a branding accessory. Solar panels made for great press releases and drone shots, but when it came to serious operations, they were rarely part of the equation. The real decisions about production schedules, curing times, plant expansions still defaulted to grid power, diesel backup, and cost-per-kilowatt logic. But that equation is shifting "Fast".

Today, renewables are no longer symbolic. They're strategic infrastructure; not just because they cut emissions, but because they stabilize energy costs, reduce procurement risk, and anchor a factory's credibility in carbon-sensitive markets. In regions where the grid is still fossil-heavy or volatile, renewables aren't just clean—they're controllable.

The best place to start? Look up.

Most precast plants have enormous roofscapes—unused real estate exposed to sunlight for 8–10 hours a day, 300+ days a year. That space is an asset. A 100 kWp rooftop solar system, for instance, can generate over 120,000 kWh annually in Southeast Asia's solar belt. That's not a side supply—that's a serious offset. When tied into net metering, feed-in-tariffs, or hybrid load balancing, it starts shaving real percentages off your annual energy bill.

And the ROI isn't theoretical. With rising electricity tariffs, PV prices falling below $0.50/W installed in many markets, and government-backed tax incentives (like Malaysia's GITA/GITE, NEM or Singapore's GreenGov S$), payback periods for solar infrastructure are dropping to 3–7 years. After that? The sun keeps producing—on weekends, holidays, and during load drops.

But the real value of renewable integration isn't just financial—it's systemic alignment.

Pair rooftop solar with smart inverters, and now you're controlling voltage profiles across your curing systems. Add a modest battery bank, and you can flatten peak demand surges that would otherwise trigger grid penalties. Connect to your PLC network, and you're able to timestamp paver curing cycles to periods of optimal solar availability by matching process energy to source timing.

This is where energy ceases to be background noise. It becomes programmable logic, and that's an engineering asset.

And while full decoupling from the grid may still be years away, the direction is clear. Every kilowatt produced on-site is one you didn't draw from a coal-fired substation. Every megawatt-hour generated is one less metric tonne of $CO_2$ in your LCA. For a product that may claim sustainability at surface level, this is the proof behind the promise.

In short: solar is no longer a bonus. It's a baseline. And the factories that embrace this shift early aren't just cleaner. They're leaner, smarter, and more likely to win the kinds of projects where carbon isn't just reported, but scored.

*"A sustainable factory doesn't just make clean products.
It makes them using clean energy, by design."*


## Final Thought

*Energy is not just a utility—it's a footprint.*

It's easy to treat energy as background noise: a cost centre, a meter reading, a technicality between production and dispatch. But in the context of sustainability, every kilowatt becomes part of the paver's identity. It leaves a trail; not just through wires and steam lines, but through carbon disclosures, EPD tables, and regulatory benchmarks.

When we understand where that energy goes, we move from passive consumption to active accountability. It's no longer about cutting costs. It's about aligning operations with values. Because a factory that can map its own energy flow is a factory that's no longer guessing at its impact—it's measuring it, managing it, and ultimately, mastering it.


## 4.4 Process Water and Recycling Systems

*Why a sustainable factory can't just manage water use—it must engineer its water intelligence.*

In the realm of concrete production, we often speak of materials and energy as the pillars of sustainability—but water, the most fundamental of resources, is frequently treated as a side concern. Not because it's unimportant, but because it has long been taken for granted.

Unlike cement, aggregates, or even electricity, water has historically arrived with near certainty: on tap, on demand, and underpriced. It moved through processes silently, used liberally, and drained away invisibly. But that era is ending.

The rise of climate volatility, urban water stress, regulatory tightening, and ESG transparency is forcing the concrete industry to confront a long-ignored truth: water is not a utility—it's a finite asset, and in manufacturing, it must be treated as such. For interlocking paver plants, which rely on water not only for batching and curing but for cleaning, cooling, and dust suppression, the implications are substantial. Because unlike structural concrete poured on-site, precast production is an intensive, repetitive cycle and every cycle pulls from the same well.

To operate sustainably, a factory must move beyond reduction. It must implement closed-loop, technically verified, process-integrated water systems; not to comply with green ratings, but to survive long-term shifts in supply, regulation, and accountability.

## Defining the True Water Lifecycle of a Paver

To understand what it means to engineer water efficiency, we must first dismantle the oversimplified notion that water use begins and ends at the mixer. In reality, the water lifecycle of a paver extends far beyond batch water.

It begins at the aggregate washing bay, where fine dusts are removed from sand and gravel; consuming thousands of litres per hour in continuous feed operations. It continues into moisture correction protocols, where pre-wetting is applied to avoid mix inconsistencies due to dry fines. Then into the mixer chamber, where precise volumes are dosed, and into curing environments, where mist systems or steam generate controlled humidity for strength gain. Beyond production, water is needed for equipment cleaning, mould washing, pigment flushing, and at the administrative periphery: toilets, cooling towers, green buffers, canteens.

In total, studies from the Concrete Sustainability Hub (MIT, 2020) and JRC Europe (2022) estimate that fully integrated precast concrete plants may use 100–180 litres of water per cubic meter of output, when accounting for both direct and indirect use.

But more critical than the total volume is this question: How much of that water must be fresh? And how much is discharged, untreated, or wasted?

## From Drain to Loop: The Shift to Engineered Water Systems

*Why the future of water in concrete manufacturing is not about using less—it's about losing nothing.*

For most of its industrial history, the concrete sector has treated water as linear: extract, use, discard. The process was straightforward. Water entered the system as a batch input or cleaning agent, flowed through the plant's operations, and exited as wastewater—often untreated, unmeasured, and unchallenged. That model is no longer defensible.

As environmental constraints tighten and the cost of inaction rises, progressive manufacturers are replacing this old paradigm with a far more robust one: closed-loop water management. This isn't about recycling as an ethical add-on—it's about reengineering the entire internal flow of water as a controlled, traceable, and optimized circuit.

The shift from drain to loop begins at the ground level. Washout pits that once collected waste slurry are now redesigned as settlement tanks, partitioned to separate heavy fines from lighter cement residues. Instead of flowing into storm drains, the greywater is redirected into filtration systems: sand beds, baffle tanks, lamella clarifiers, or geotextile cartridges. The aim isn't just to clean, but to requalify. Water isn't leaving the system; it's changing roles, shifting from runoff to input, from discharge to asset.

From here, treated water is fed back into mixing lines, curing sprayers, or cleaning loops. Some systems incorporate inline pH control using $CO_2$ bubbling or buffering agents like citric acid or soda ash, adjusting the high alkalinity of cementitious water to within safe reuse margins. At each stage, sensors monitor turbidity, conductivity, and chemical stability—not as afterthoughts, but as part of the same QA logic applied to aggregates and binders.

In the most advanced facilities, this system is fully automated. Pumps and valves open or close based on water quality readings. If turbidity is too high, it's routed back for re-settlement. If conductivity spikes, dilution from fresh supply is triggered. These systems don't just respond—they learn. Over time, they establish patterns and thresholds that can be used to calibrate operations, detect anomalies, or even inform predictive maintenance.

The result? A production process where water doesn't exit the system prematurely. It loops. It transforms. It returns to work. And in doing so, it becomes part of the factory's efficiency, not its environmental debt.

More importantly, this isn't a niche practice anymore. Across Europe, especially in Germany and the Netherlands, precast plants with over 80% water recovery are the norm—not the exception. In Malaysia, the Department of Environment (DOE) is increasingly linking discharge licenses to proof of internal recycling systems, with structured water reuse targets embedded in Green Building Index (GBI) certifications. Singapore's BCA Green Mark similarly awards significant points for closed-loop water reuse in concrete production.

This isn't innovation for its own sake. It's the logical evolution of an industry under pressure—not just to do better, but to prove it.

*"A factory that loops its water doesn't just lower its usage.*
*It rewrites its responsibility."*

## Concrete Performance with Recycled Water: Myths and Material Science

*Why water reuse isn't a compromise—it's a matter of control.*

Ask any concrete producer why they're hesitant to use recycled water, and you'll often hear the same concerns: "It'll affect the strength." "We don't know what's in it." "It's not worth the risk." These hesitations are rooted not in poor logic, but in a legacy mindset—one that shaped by decades of habit, liability, and a preference for the known quantity of fresh, potable water.

But engineering doesn't move forward by fear. It moves forward by measurement.

And over the past two decades, the measurement has become increasingly clear: recycled water, when properly treated, monitored, and integrated can meet or exceed the performance demands of standard concrete production, including interlocking paver manufacturing. The issue has never been the recycled water itself. It has always been the lack of control around it.

Let's break that down.

Water used in concrete doesn't simply lubricate a mix. It participates in hydration chemistry, affects workability, controls setting time, and influences curing dynamics. When water is dirty or contains sugars, organic contaminants, high levels of salts or unreacted fines; it can and will disrupt this balance. But when its properties are known, controlled, and maintained within technical boundaries, recycled water becomes just another engineered input.

The global standards already acknowledge this. ASTM C1602, EN 1008, and IS 456 all allow for the use of non-potable or recycled water in concrete manufacturing, provided that certain quality thresholds are met. These include:

- pH levels typically between 6.0 to 8.5
- Limits on chloride (≤ 500 ppm) and sulphate (≤ 1000 ppm) content
- Total suspended solids (TSS) ideally below 50,000 mg/L, with tighter limits for decorative or exposed-surface products
- Absence of harmful organic substances, such as sugars, oils, or reactive admixture residues

These are not arbitrary numbers. They reflect empirical thresholds at which hydration behaviour, setting kinetics, and long-term strength begin to deviate from control mixes. But they also offer an expansive envelope for safe, predictable use of treated process water.

In practice, most recycled water used in precast plants is a blend of:
- Mixer washout water (cement-rich, alkaline, high in fines)
- Curing system runoff (often clean or lightly contaminated)
- Surface cleaning water (moderate turbidity, low in chemical load)

When this water is passed through multi-stage settlement tanks, filtration units, and pH balancing systems, it typically returns to acceptable thresholds. More importantly, consistent reuse within the same process (i.e., same mix design, same hydration window) introduces repeatable conditions, which is one of the most critical elements in quality assurance.

In well-documented case studies from Germany, Japan, and Singapore, recycled water has been shown to maintain:
- Equivalent 28-day compressive strengths (within ±3% of control)
- No measurable increase in shrinkage or creep for GGBS-rich mixes
- Consistent slump performance, especially when SCMs are present to buffer pH variability
- Stable setting time, provided admixture interactions are understood and accounted for

In fact, a 2021 study in *Cement and Concrete Composites* demonstrated that mixes incorporating up to 80% recycled process water exhibited improved particle packing and paste density, and this likely due to the presence of ultra-fines and nucleation sites in the washout water.

## What does this mean for interlocking pavers?

It means that recycled water can and does work; not just for base layers or low-grade blocks, but for high-density, high-durability, tight-tolerance pavers. The key is integration:

- Pre-treat the water, don't just reuse it
- Monitor it daily as part of QA, not weekly as an afterthought
- Adjust mix designs slightly, where necessary, to account for fines content or pH buffering
- Blend with fresh water if conditions shift (e.g., during peak evaporation seasons or after equipment overhauls)

What this unlocks is not just sustainability. It unlocks resilience. In markets where water access is strained, either by drought, municipal restriction, or cost; plants that manage their own internal water loop hold a distinct operational advantage.

*Recycled water doesn't weaken concrete. Mismanaged water does. The difference is not in the source—it's in the control.*

## Rainwater Harvesting, Dewatering, and Slurry Recovery: Extending the Loop

Water sustainability doesn't end with internal recycling—it extends to external capture and post-use recovery.

Rainwater, when properly harnessed, can provide a free, low-impact supplement to non-potable demand. Roof catchment systems feeding to underground cisterns, filtered through leaf separators and sand beds, have proven capable of meeting 30–40% of curing water requirements during monsoon seasons in India, Malaysia, and Indonesia. In places like Singapore, BCA Green Mark-certified precast facilities routinely integrate rain harvesting into their certification pathways.

But more complex and more valuable than water recovery is slurry recovery. Traditionally, washout pits filled with cement-laden water were left to settle, dry, and be scooped out as solid waste. But modern factories are now installing filter press systems that separate solids and liquids at high efficiency. The solids which are generally rich in fines safely can be reused in low-grade block mixes, road base, or sent to third-party recyclers. The water? Clarified, neutralized, and re-fed into curing systems or used for equipment cleaning.

Some have gone even further, capturing overspray from mist curing systems via stainless-steel trays, channelling the water back into secondary holding tanks. What once evaporated now returns to the loop—closing a loss that few factories even thought to measure.

*"What you don't count, you don't control. But what you recover, you reclaim—environmentally and economically."*

## Final Thought

*When you follow the water, you find the truth about how your factory thinks.*

Water is more than a raw material in concrete production—it's a mirror. It reflects how carefully a system is designed, how well operations are understood, and how seriously a manufacturer takes the difference between convenience and responsibility. It doesn't matter how green a paver claims to be if it's made in a factory that treats water as disposable. That kind of sustainability is cosmetic.

The factories that stand apart today, and will still be standing strong tomorrow are the ones that don't wait for water to become a crisis. They anticipate it as a constraint, plan for it as a resource, and engineer their systems to lose as little as possible. Whether it's rain from the roof, slurry from a washout, or the mist that settles after curing—it's all material. All are measurable...All are meaningful.

And in the growing movement toward environmental product declarations, circular manufacturing, and net-positive production models, those who can account for every drop will hold every advantage.

*"A sustainable product begins not with what you use,*
*but with what you no longer waste."*

## 4.5 Waste Minimization and Raw Material Efficiency

*Why sustainability isn't only about what goes in—but about what doesn't come out.*

In the industrial world, waste is often seen as inevitable; usually as an accepted by-product of volume, speed, and complexity. But in modern concrete manufacturing, and particularly in the high-frequency world of interlocking paver production, that mindset no longer holds. Waste is no longer a by-product. It is a performance indicator; it's a direct signal of inefficiency, poor planning, or unclosed loops. And in a climate-aware manufacturing environment, inefficiency is no longer tolerable, let alone invisible.

The raw materials involved in concrete production—cement, aggregates, SCMs, water, pigments, and admixtures—are not just commodities. Each comes with an extraction footprint, an energy cost, a logistics burden, and an emissions profile. If they're wasted, whether through process loss, overdesign, mishandling, or lack of precision; they don't just reduce profit margins. They add unnecessary carbon to every square meter of paver you produce.

That's why the most forward-looking factories are now treating waste minimization and material efficiency not as CSR buzzwords, but as engineering disciplines. They design workflows and equipment layouts that actively prevent waste, install feedback systems that capture off-spec loss in real-time, and embed reuse pathways into the very architecture of their operations.

## Understanding the Full Waste Profile of a Paver Plant

Walk through any interlocking paver plant on a busy day and you'll likely miss most of the waste. It doesn't always look like failure. It doesn't pile up dramatically. It doesn't always get measured. That's because in high-throughput operations, waste hides in fractions: half a bucket of pigment discarded during changeover, slurry from a rushed washout, a few rejected pallets tucked behind the batching bay, overspray mist lost to the wind. Individually, they seem negligible. Collectively, they write the story of resource loss.

Understanding waste in paver production begins with one mental shift: waste is not just broken pavers or spilled aggregates. It's anything that leaves the production system without delivering value. And once you adopt that lens, the picture broadens rapidly.

Consider the batching stage. Every mis-calibrated scale, every moisture correction missed, every manually overridden admixture dose results in a batch that may technically meet strength, but consumes more cement, water, or pigment than needed. These aren't rejected units. They're overdesigned outcomes, burning through carbon and cost invisibly.

Downstream, poor demoulding timing, especially under aggressive curing regimes leads to microcracks, chipped corners, or uneven finish. These pavers are often culled, stacked aside, or sent for reprocessing. But their material cost is already paid.

Slurry pits, if unmanaged, overflow with high-alkaline wastewater, which usually contains cement, fines, and fibres suspended in discharge. If not reclaimed, this becomes not just a waste disposal problem, but a regulatory one. Pigments flushed prematurely, aggregates spilled during transfer, untracked shrinkage or deformation during curing—each tells the same story: material that entered the gate with cost, but exited with none of its value realized.

In short, the waste profile of a paver plant is multidimensional:

- *Material waste* – cement, aggregates, water, admixtures, and pigment lost through overdesign or process inefficiencies
- *Product waste* – cracked, deformed, chipped, or off-colour pavers rejected before dispatch
- *Process waste* – time lost, energy used inefficiently, or storage overflows due to poor inventory alignment
- *Chemical waste* – unutilized admixtures, effluent with residual solids, high-pH rinsewater
- *Operational loss* – downtime, mis-batching, or failed QC that forces entire runs into low-grade reuse

And the problem isn't just the waste itself. It's that very little of it is routinely mapped, logged, or used for improvement. Without visibility, accountability becomes guesswork. Without metrics, sustainability becomes rhetoric.

In recent lifecycle mapping efforts across multiple Southeast Asian plants, waste from concrete production was found to go as high as 7% (maybe even more) of total raw material intake, and guess what, much of it avoidable. And yet, only a fraction of facilities had continuous tracking systems in place to quantify it in real time.

The takeaway is clear: the full waste profile of a plant cannot be understood by looking at the skips at the back. It must be engineered into visibility; through measurement, mapping, and cultural alignment. Because in the language of efficiency, what you don't track, you silently accept.

*"Waste isn't just the material that leaves the system.*
*It's the opportunity that never got used."*

## Designing for Less: Material Efficiency Starts Upstream

By the time waste reaches the factory floor, it's already too late. Material has been ordered, transported, unloaded, and often processed before it ever has the chance to be wasted. That's why true efficiency isn't about minimizing what's thrown away; it's about not giving waste a chance to begin. And this shift doesn't happen in the skip bins or slurry pits. It happens upstream, at the design table, in the software that generates the mix design, and in the decisions made before the first batch is poured.

Upstream material efficiency begins with understanding the role of precision in manufacturing. In the concrete world, imprecision isn't always obvious. A dosing system that over-delivers cement by 1% per mix may still produce a strong, high-quality paver—but over time, that small excess compounds into tonnes of overuse. The same goes for admixtures that are added conservatively "just in case," or pigment loads bumped up to ensure colour uniformity under variable curing. These are not errors. They are habits of overdesign; built for safety, not efficiency.

Modern factories are breaking this pattern by building control into the front-end. High-resolution weigh scales and real-time moisture sensors allow for tighter batching tolerances. Precision flow meters ensure water-cement ratios are accurate to the decimal. Automated pigment dispensers dose based on cumulative cycle logs, reducing over-pigmentation that leads to costly rework or aesthetic rejection.

But the most forward-looking efficiency gains come not from equipment, but from intentional design of the product itself. Factories are revisiting paver geometry, edge profile, and interlock configuration to reduce:

- Material consumption per unit
- Edge loss during compaction or demoulding
- Stress concentrations that lead to surface chipping or hairline cracks
- Overreliance on cement-rich outer layers for abrasion resistance

In one leading European plant, a redesign of a 400x400mm paver series with optimized corner radii, thickness tapering, and core voiding—reduced raw material usage by 9.2% per unit without compromising strength, load distribution, or dimensional stability. That's not waste reduction after the fact. That's prevention by design.

Even at the mix design level, digital simulation tools now allow for advanced modelling of:

- SCM substitution curves
- Aggregate packing efficiency
- Workability flow under temperature variance
- Early vs. long-term strength optimisation
- Shrinkage potential based on raw material chemistry

This allows engineers to formulate mixes that meet performance targets precisely, rather than exceeding them to mask unknowns. The result is not just less waste, but also less conservatism in formulation, leading to lower embodied carbon, better cost control, and fewer rejections due to overcompensated inputs.

And upstream doesn't stop at formulation. It extends into procurement alignment and stock planning. Batching the right quantity, with the right volume buffer, reduces return loads, overrun product, and the temptation to run "just-in-case" batches that overfill yard capacity. Well-run plants now integrate ERP data with real-time production logs, ensuring materials are only released when orders, moulds, and curing cycles are confirmed—not when assumptions are made.

*"Upstream efficiency isn't about cutting corners—it's about refusing to guess. It's about designing exactly what's needed, and trusting the process to do only what's necessary."*

In an industry where margins are tight and carbon audits are growing sharper, this level of design precision is not luxury. It's survival.

## Reclaiming Internal Waste: A Closed-Loop Mindset

Even in the most optimized paver production systems, not all waste can be prevented. Machinery malfunctions. Mixes get rejected. Pigments bleed, edges chip, batches run short. But the difference between a plant that tolerates loss and one that operates sustainably is what it does next. In a circular production mindset, waste is not an endpoint. It's an asset waiting for reclassification.

The principle is simple: if the material has already entered the factory, it should stay in play as long as possible. That doesn't mean dumping broken pavers back into a new mix. It means developing controlled, validated, and technically sound pathways for reintroducing that material—intelligently and without performance compromise.

This is where a closed-loop waste recovery system starts to become a value-generating process, not just a damage-control mechanism.

Take rejected or off-spec pavers. Instead of sending them straight to landfill, many advanced plants now run them through dedicated crushing units; recovering the coarse fraction as recycled aggregate. When size-graded and free from coatings or contaminants, this material can be used:

- In subbase layers beneath yard pavements
- As a partial replacement for virgin aggregate in non-architectural blocks
- As an input in blended concrete for kerbs, spacers, or non-load-bearing units

But success here requires more than good intentions. It demands clear technical limits. Excessive fines, inconsistent gradation, or variable absorption can introduce durability risks. That's why closed-loop reprocessing is best done internally, where quality is known, and integration points are controlled.

Plants that succeed treat crushed returns not as second-tier material, but as a documented input with its own spec.

Similarly, washout sludge which once considered a problem, now can be dewatered and recovered. The fines extracted (typically rich in un-hydrated cement particles and fine sand) are now being used in:

- Base material for road access within plant boundaries
- Filler blends for extruded concrete pipes
- Shotcrete mixes with modified polymer content

Even coloured washout water, when filtered to remove coarse pigment residue, can be reused in colour-consistent production cycles; minimizing pigment waste without risking chromatic drift. Some plants tag batches by water source, ensuring that minor tone shifts are consistent within each production lot.

The most progressive factories go a step further: process-integrated reuse loops. Here, waste isn't transported, stockpiled, or repurposed days later. It's intercepted midstream:

- Pigment tanks are flushed into a holding loop, reused in dark-colour runs
- Short-pour leftovers are moulded into custom-size fill blocks for in-house use
- Excess batch water is directed to non-structural spraying applications
- Dust from filter systems is compacted and blended into base paver mixes

This kind of circularity isn't accidental. It's the result of systems thinking, where every input has a second life engineered into the process. In a full-loop plant, the disposal bin becomes a planning failure—not an inevitability.

And the benefits go far beyond material savings. They extend to:
- Reduced landfill tipping fees and regulatory disposal costs
- Lower virgin material demand, especially for coarse aggregates
- Improved EPD scores, where reclaimed content counts toward embodied carbon targets
- Higher resilience, especially in remote or supply-constrained environments

*"Circularity in production isn't about lowering your standards. It's about raising your system's intelligence."*

The result is a factory that doesn't just produce concrete—it recycles certainty. Every recovery pathway becomes part of its performance promise. Every bypassed bin becomes a measure of progress.

## Measurement = Accountability

Factories rarely improve what they refuse to measure. And in paver production, where inefficiencies often manifest in grams per unit or seconds per cycle, the absence of measurement doesn't just mean missed savings—it means missed truths. Because every claim of sustainability, material efficiency, or closed-loop manufacturing rests on a foundation that's either validated… or vague.

That's why the modern concrete plant must treat measurement not as admin work, but as a design tool. A waste-minimizing operation doesn't guess how much cement it uses per $m^2$. It calculates it. It doesn't estimate pigment loss. It logs it. It doesn't rely on intuition to reduce rejections. It tracks cause-specific rejection ratios, over time, by operator and machine.

To build real accountability, data must be embedded into the flow of operations, and not retrofitted as an afterthought.

What Should Be Measured?

There is no single metric for material efficiency. Instead, the factories that succeed in reducing waste maintain a multi-layered data structure. They track:

- **Material input-to-output ratios:** How much cement, aggregate, water, pigment, and admixture go in per unit—and how much product is actually delivered?
- **Rejection rates:** Not just as totals, but categorized—chipped corners, colour variation, dimensional mismatch, surface pitting, premature curing, demoulding cracks.
- **Yield per batch:** How does theoretical yield compare to actual mould output? Are moulds being overfilled? Are batch volumes being adjusted without feedback?
- **Recycled content reintegration:** How much material (crushed pavers, slurry fines, returned water) re-enters the system, and in which products?
- **Process waste:** How much water is lost between cycles? How much pigment is flushed during colour transitions? How much dust is lost during dry loading?

In advanced plants, this data is integrated into digital dashboards with batch-specific records, real-time alerts, and cross-linked QA tracking. These aren't just tools for plant managers. They're systems that connect the operator's decisions to the factory's carbon footprint.


## From Data to Action

But measurement alone is not accountability. Data becomes meaningful only when it changes behaviour.

That's why many manufacturers now incorporate material efficiency KPIs directly into team performance reviews. Line operators are rewarded not just for hitting output targets, but for reducing rejections, minimizing overuse, and improving recovery rates. Engineering teams are tasked with optimizing mix designs to use less, not just perform more. Even procurement departments are asked to forecast smarter to avoid overstocking that leads to expiry-based disposal.

*"In high-performance manufacturing, measurement isn't just record-keeping. It's direction-setting."*

Measurement is the bridge between intent and execution. It's what separates aspirational sustainability from operational precision. Because in a world of carbon benchmarks, procurement scoring, and third-party certifications, what you can't prove, you can't claim. And what you can't measure, you'll never control.

## Final Thought

*Every paver carries a story—not just of what it's made of, but of what was wasted in making it.*

Sustainability in manufacturing doesn't start with bold claims or green labels. It starts with restraint. It starts with asking: did we use more than we needed? Did we waste what we could have reused? Did we design systems to protect our materials or did we rely on habit and hope?

In the interlocking paver industry, where volumes are high and margins tight, efficiency is not a matter of polish; it is in fact a matter of ethics and engineering. To minimize waste is not simply to clean up better; it is to think better. To treat every kilogram of raw input as something valuable. Something earned. Something with an ecological cost that must be honoured, not squandered.

The factories that will lead in the next decade are not those that boast the highest output. They're the ones that extract the most value from every tonne they bring in. And leave behind the least in their wake.

## 4.6 Power and Purity: Rethinking What Enters and What Lingers

*Why sustainability means asking not just how clean the process is—but how safe the product remains.*


### 1. Cleaner Fuels and Renewable Integration: Moving Beyond Fossil Fuel Dependency

When we think about reducing environmental impact in manufacturing, the most immediate concern is energy. The manufacturing sector, including concrete plants, has long been powered by fossil fuels—coal, gas, diesel—all of which contribute heavily to carbon emissions and, ultimately, climate change. The solution seems simple: transition to cleaner, renewable sources. But that shift isn't merely about swapping one fuel for another. It requires a complete rethinking of energy use within the plant and, most importantly, how energy flows through the production process itself.

Historically, fossil fuels have been the default energy source for processes like curing, mixing, and heating. In paver plants, for example, gas-fired boilers and diesel-powered generators fuel steam curing chambers that operate at temperatures up to 80°C. While these systems have proven reliable and cost-effective, their carbon footprint is undeniable. But clean alternatives are emerging, such as solar power, wind energy, and electric heat pumps, and they're not just theoretical. They're becoming operational.

In the last few years, some of the most energy-conscious plants have begun adopting photovoltaic (solar) panels on their rooftops to offset grid power consumption. In fact, a recent project in one factory demonstrated that a 250-kW solar array installed on the factory roof could meet 25–30% of the factory's energy needs during peak daylight hours—This cut both energy costs and carbon emissions.

Similarly, the integration of heat pump technology has begun to replace traditional gas-fired boilers in certain curing systems. These electric systems transfer heat rather than generate it from combustion, offering much higher efficiencies, sometimes could be up to 300% energy efficiency with no direct carbon emissions. In Northern Europe, plants have also leveraged district heating systems, where excess energy from nearby industries is used to provide the thermal energy needed for curing pavers.

But renewable energy is not limited to production processes. Some plants are considering energy storage systems, such as lithium-ion batteries, to store excess energy generated by solar panels during the day, which can then be used during peak operational hours, further reducing reliance on grid energy. By incorporating both active generation (solar, wind) and energy storage systems, the factory moves toward a fully integrated, closed-loop energy solution.

*"Shifting to cleaner fuels isn't just about being green. It's about decoupling from fossil fuel dependency and gaining autonomy in how power is sourced and consumed."*

## 2. Toxicity and Lifecycle Safety: From Material Sourcing to End-of-Life

The second part of the sustainability challenge lies not in the energy used to make the paver, but in the chemistry of the paver itself.

Concrete, by its nature, is durable, versatile, and long-lasting. But while its strength is its greatest asset, it can also be its most dangerous feature if the wrong materials are chosen, or worse, if those materials carry harmful chemicals that remain within the product throughout its lifecycle. What happens before the paver is installed matters just as much as what happens when it's left exposed to sun, rain, or foot traffic for decades.

Consider the pigments used to colour pavers. Many manufacturers still rely on heavy metals such as chromium, cadmium, or lead to achieve certain hues. While these metals offer vibrant colours, they are toxic and can leach into the environment over time, particularly when exposed to moisture. It's a common oversight: a product designed for outdoor use, intended to withstand the elements, may also be leaching toxins into stormwater, potentially contaminating local ecosystems and waterways.

This same issue extends to certain admixtures and surface coatings. Some curing accelerators or finishers; especially older formulations contain volatile organic compounds (VOCs), which can contribute to air pollution or cause health hazards to workers and end users. Even the cementitious binders used in concrete can present issues: while modern formulations of fly ash and GGBS reduce environmental impact, traditional Ordinary Portland Cement (OPC) can contain traces of toxic elements like arsenic, mercury, and benzene.

This isn't a minor issue. The tens of thousands of tonnes of pavers produced globally every day could contain hazardous materials that pose a threat during manufacturing, installation, and eventual disposal. As cities become more densely populated, and as the world becomes more conscious of toxic legacy materials, these concerns can no longer be ignored. Sustainability is not just about carbon; it is about chemical safety and long-term human health.

## Designing for Non-Toxicity

The solution to the toxicity problem lies in making informed material choices. For example, pavers can be coloured with iron oxide-based pigments, which are non-toxic, stable, and highly durable. Similarly, eco-friendly admixtures like bio-based accelerators can replace their chemical counterparts, offering the same performance without VOC emissions or harmful residues. Even cement formulations are evolving to use lower-clinker blends, which not only reduce the carbon footprint but also minimize the inclusion of harmful trace elements.

But toxicity control doesn't stop with the mix. Recycling is another critical component. When pavers are decommissioned at the end of their lifespan whether due to damage, redesign, or repaving; it's essential that the raw materials are safe to recycle. Heavy metal contamination or embedded chemicals can render a paver unsuitable for reuse in other products or applications, perpetuating the environmental cycle of waste rather than closing it.

## The Global Shift Toward Non-Toxic, Low-Impact Products

This integrated thinking is gaining traction. In the United States and Europe, numerous certifications now address product safety and material transparency, including Declare labels, Cradle to Cradle (C2C), and LEED v4.1 credits. Green certifications are increasingly requiring manufacturers to disclose material safety data, and municipalities are adopting stringent sustainability codes that limit the use of toxic substances in public infrastructure.

Countries like Germany, which already have stringent building materials regulations, are leading the charge on sustainable construction by mandating low-VOC, heavy-metal-free, and low-carbon construction materials. Meanwhile, the European Union's REACH regulations (Registration, Evaluation, Authorization, and Restriction of Chemicals) set the standard for chemical safety, banning many harmful substances and driving material manufacturers to innovate with safer alternatives.

In Malaysia and other rapidly developing regions, environmental performance and product safety are starting to be non-negotiable. As regulatory bodies tighten oversight on material composition, plants will face pressure to stay ahead of these trends—not just to improve market share, but to stay compliant.

*"Sustainability doesn't end at the production gate. It extends through every phase—design, installation, life, and end-of-life. A sustainable product starts by thinking about what it leaves behind."*

# 4.7 Certification, Traceability, and Plant-Level EPDs
*Why proving sustainability matters just as much as practicing it.*

In the early days of sustainable construction, a company's commitment to the environment was largely judged by intention. A solar panel on the roof, a few low-energy lights in the office, a "green" product line in the catalogue; and these were enough to position a business as eco-conscious. Today, that era is over.

Sustainability is now **Measurable. Reportable. Auditable**. And in many procurement scenarios; especially for infrastructure, government, and commercial projects, it is increasingly contractual. If a product claims low carbon, it must show an EPD. If a plant says it recycles water, it must show the system logs. If a manufacturer claims low toxicity, it must produce test results or third-party validation. This is not a burden. It's a new standard. And for interlocking paver manufacturers, it's an opportunity to turn operational discipline into market advantage.

## Certification Systems: Navigating the Alphabet Soup
The world of environmental certification is crowded, often confusing, and constantly evolving. But at its core, the goal is simple: to provide trustworthy benchmarks for product sustainability. Here are the most relevant ones for concrete pavers and precast operations:

### 1. LEED (Leadership in Energy and Environmental Design) 1
Administered by the U.S. Green Building Council, LEED remains one of the most recognized green building frameworks globally. In LEED v4.1, concrete pavers can contribute to credits under:

- *Heat Island Reduction* (via high SRI values)
- *Construction and Demolition Waste Management* (if made with recycled content)
- *Material Ingredient Reporting* (if part of C2C program)
- *Environmental Product Declarations* (under the Materials & Resources category)

## 2. MyHIJAU (Malaysia)

Managed by the Malaysian Green Technology and Climate Change Corporation (MGTC), MyHIJAU provides official recognition for environmentally friendly products and services in the Malaysian market. Paver manufacturers with energy-saving production methods, material reuse, and low-toxicity formulations can qualify for listing, which is essential for eligibility in green public procurement (GPP) projects.

## 3. GreenRE (Malaysia)

Primarily used in private-sector developments, GreenRE includes criteria for construction material performance, thermal comfort, and embodied energy. Precast pavers with validated EPDs or that use recycled aggregates or SCMs often contribute to multiple credit pathways.

## 4. Declare Label / Cradle to Cradle (C2C)

Declare focuses on material transparency, requiring manufacturers to list every ingredient in their product, right down to 100 ppm, along with associated health impacts. C2C certification goes further, evaluating material health, reutilization, renewable energy use, and social fairness.

*"In the era of greenwashing fatigue,*
*independent third-party certification becomes not just a badge;*
*but a filter for serious buyers."*

## Traceability: From Batch to Block

Beyond certification, the next frontier in sustainable concrete production is traceability, which is the ability to track every paver back to its inputs, process conditions, and performance characteristics. This is where operations become not only transparent but verifiable.

Progressive plants now incorporate digital traceability systems that:

- Assign batch numbers to every pallet of pavers
- Link each batch to a mix design version, including source of cement, aggregates, admixtures, and pigments
- Capture curing temperature profiles, water content, and production energy usage
- Store QA/QC data, including compressive strength, dimensional checks, and surface finish assessments

This level of granularity isn't just internal. It enables project-level material documentation, where contractors or developers can verify that:

- A specific lot of pavers used for a green-certified project came from a recycled mix
- The water used in that production week met reuse targets
- The SRI rating of a surface meets a design specification under a LEED credit

In Europe, traceability is now being formalized through Digital Product Passports (DPPs) under the EU Green Deal. These tools aim to encode sustainability attributes into a scannable record that stays with the product across its lifecycle—from factory to job site to eventual reuse or recycling.

## Plant-Level EPDs: Turning Data into Competitive Edge

At the heart of this accountability system is the Environmental Product Declaration (EPD)—a standardized, third-party verified document that quantifies a product's environmental impact over its entire lifecycle.

For interlocking pavers, an EPD typically includes:
- Embodied carbon (GWP) from raw material extraction through manufacturing
- Water consumption (total and recycled)
- Energy usage (by type: renewable, non-renewable)
- Waste generation
- Acidification, eutrophication, and ozone depletion potential

EPDs follow international standards such as ISO 14025 and EN 15804, and are based on Product Category Rules (PCRs) specific to construction materials. These documents do not claim a product is "green" or "certified." Instead, they report objective impact data, which allows buyers, architects, and regulators to make side-by-side comparisons between different products.

Why does this matter?

Because in carbon-regulated markets, EPDs are fast becoming required documentation, and not just a value-add. In Sweden, Germany, and the Netherlands, public procurement projects now score bids partially based on the embodied carbon profile of specified materials. In the UK, PAS 2080 encourages infrastructure providers to select lower-carbon materials based on comparative EPDs. Even in Southeast Asia, we're seeing early signals of this trend, particularly in transit, commercial development, and data centre design.

Having a plant-specific EPD, rather than a generic or association-level one; provides distinct advantages:

- It reflects your actual operational practices, including energy source, water recycling, and material recovery
- It allows you to show improvement over time, especially if you're investing in low-carbon cement, renewable integration, or digital batching
- It provides a credible basis for carbon offsetting or neutrality claims, when coupled with external verification frameworks

*"In a world moving toward quantified carbon, an EPD isn't just paperwork. It's proof of engineering precision and a passport into premium markets."*

## The Future: Verified Products in Verified Systems

As the industry matures, we are headed toward a future where every building product; that includes every block, slab, or paver, will be accompanied by data. Not because it's trendy, but because supply chain decarbonization, ESG compliance, and lifecycle analysis demand it.

And the plants that adapt early by investing in measurement systems, engaging with credible certification bodies, and designing for traceability, will not only lead in environmental performance. They'll own the trust that clients, contractors, and cities are increasingly seeking.

Because in the next era of construction, your process is your product.

*"If you can't show how it was made, how do you expect anyone to trust what it claims to be."*

## 4.8 Chapter Reflection

*In a truly sustainable system, nothing is hidden—not the inputs, not the impacts, and certainly not the intentions.*

Chapter 4 began with a question: how can a product claim sustainability if its making is unchecked, its emissions unknown, its ingredients unverified? We've answered that by going deep into mix chemistry, energy loops, water cycles, recovery systems, traceability tools, and the silent threats of toxicity. And in doing so, we've dismantled the idea that sustainability is a surface claim. It is a discipline. A system of decisions. A chain of accountability from quarry to quality log.

Every paver that leaves a factory carries with it not just weight and shape, but a history. If that history is wasteful, toxic, or unverifiable, no amount of green branding can make it right. But when it's built in a factory that manages energy precisely, closes its water loop, reclaims its materials, and publishes its data, it becomes something else entirely: a transparent building block of a better city.

**In an industry where carbon, compliance, and performance are no longer negotiable—only the honest product wins.**

# Chapter 5: Product Durability and Lifecycle Performance

*Why long-term sustainability begins with long-term strength.*

Sustainability isn't just about what goes into a product. It's also about how long the product stays out of the waste stream.

You can make concrete pavers with low-carbon cement, reclaimed aggregates, solar-powered curing, and recycled water; but if it begins to crack after five years in service, the environmental benefits collapse with it. Every replacement restarts the cycle: more extraction, more mixing, more energy, more emissions. In that moment, the original sustainability promise is lost; not because of bad intent, but because of short-sighted engineering.

Durability, then, is not a bonus feature. It is the backbone of sustainability. It is the measure of whether a product truly delivers the value that it promises; not just in the moment of specification, but across decades of use. It asks:

- Can this surface endure repeated loading, without loss of interlock?
- Will it resist UV degradation, algae, and abrasion in open environments?
- Is it engineered to withstand thermal movement, freeze–thaw cycles, and heavy rainfall—not just once, but thousands of times?

And equally important: What lessons do we learn from failure? When pavers fail early due to crack, shift, discolour, or wear unevenly; what does that tells us about the original assumptions in design, material choice, installation method, or maintenance planning?

In this chapter, we examine the science, strategies, and silent risks behind long-term paver performance:

- We'll unpack the mechanics of stress and fatigue in modular systems
- Explore how weather, water, and wear break down even the strongest surface
- Dissect failure modes as diagnostic tools, not just problems
- And consider how smart design and planned maintenance can turn decades into a lifespan, not a liability

Because in the language of environmental responsibility, the greenest surface isn't the one that recycles well. It's the one that *doesn't need to* recycled for the next 30 years or more.


## 5.1. What Does Durability Really Mean in Practice?
*Why strength alone doesn't guarantee a long life.*

In engineering documents, durability often hides behind numbers, typically compressive strength, abrasion resistance, water absorption, freeze–thaw cycles. But in the real world, durability is not a test result. It's a lived outcome.

Durability is how a surface behaves on the fifth monsoon season, not how it tested on day 28. It's the way a public plaza still holds form after a decade of daily foot traffic, or how a loading bay endures thousands of wheel rotations without shifting or cracking. It's not the lab result; it should be the lifespan measurement.

And yet, in design and procurement conversations, durability is often collapsed into a single metric, that is the compressive strength. But strength, while necessary, is not sufficient.

In fact, many of the most premature paver failures happen not because the concrete was weak, but because:

- The jointing failed under lateral pressure
- The base was improperly compacted

- The edge restraints lost integrity
- The surface deteriorated under UV, algae, or freeze–thaw cycling
- Or the design didn't match the application

Durability, in this sense, is a system property. It's not just what the paver is made of—it's how that unit performs in context: under load, over time, with real users, in real weather, through changing seasons.

## Defining Durability: Beyond Strength
*Because what fails first is rarely the concrete itself.*

Durability, in the world of interlocking pavers, has often been reduced to a concrete mix specification. If the compressive strength exceeds 50 MPa and the water absorption is below 5%, the product is assumed to be durable. But that's a narrow lens, and a dangerous one too.

A paver can pass every lab test and still fail in the field. Because true durability isn't embedded in one material property. It emerges from the interplay between materials, design, and installation. It's not about resisting failure once; it is about managing the friction of use over time.

So what does real durability look like?

It starts with structural performance. The paver must resist fracture under static loads and shear along interlocking faces. It must absorb repetitive pressure from foot traffic, wheels, pallets, and weather. But that's just the skeleton.

Durability also demands dimensional stability. Pavers must sit flat. They must not rock, shift, or sink. The moment that alignment is lost, the system begins to unravel which leads to trip hazard, unevenly distributed load, and surface degradation accelerates.

Then there's surface integrity; a concept often ignored until the wear shows up. Can the paver resist abrasion from sand-laden shoes? Can it endure bicycle tyres, shopping carts, forklifts, and sweeping machines? Does it hold colour against UV? Or does it slowly become a pale, pitted echo of its original state?

And what of environmental resistance? A durable paver lives outdoors. That means it must survive cycles of wetting and drying, exposure to acids in rain or cleaning products, freeze–thaw expansion, moss and algae growth in shaded corners, or salt from de-icing efforts in cold climates. Durability doesn't stop at the edge of the block. It continues into what lives on and around the surface.

But perhaps most critically, durability is systemic. A perfect paver, installed without proper joints, on a poorly drained base, with no edge restraint, is already on a timer. The system must be in harmony:

- The subbase must support
- The joints must transfer
- The layout must distribute
- The edges must restrain
- The water must leave the surface without lingering

If even one part of that choreography breaks, the surface will degrade—not because the product was bad, but because the system was unbalanced.

***Durability isn't about what's strongest. It's about what stays together longest—under pressure, through neglect, across seasons, and in spite of everything.***

## Service Life vs. Design Life

*Because what's promised in a spec sheet isn't always what survives in the street.*

The term "design life" shows up in almost every product brochure. It's tidy, reassuring, and conveniently abstract—20 years, 25 years, 30 years. But out in the field, design life is less of a guarantee and more of a theory. It assumes the installation was perfect, the loads were predictable, the weather was average, and maintenance happened on time. In other words, it assumes a world that doesn't exist. Service life is what happens instead.

Service life is the truth-teller. It's the period during which a surface continues to perform, both structurally and aesthetically without needing repair, rework, or replacement. And unlike design life, it is shaped not just by engineering, but by human behaviour, environmental unpredictability, and practical limitations on maintenance.

A paver may be designed to last 30 years, but if joint sand is lost in year two and never replaced, failure can begin by year five. If drainage is ignored and water pools along an edge, freeze–thaw stress can fracture the surface in a single winter. If the same turning radius is used by forklifts in a warehouse courtyard every day, rutting or shoving can occur long before the mix reaches its predicted fatigue threshold.

In other words, the gap between design life and service life is where most sustainability claims quietly fall apart.

That's why forward-thinking engineers and manufacturers no longer speak in isolated life expectancies. They focus on:

- How the surface interacts with use
- What failure modes are most likely in that context
- Whether those failures are slow and repairable, or fast and systemic

- And how the product behaves when imperfectly maintained, which is the only realistic scenario for most public or commercial applications

*"True durability isn't the ability to endure ideal conditions. It's the resilience to perform despite imperfect ones."*

This is where lifecycle thinking moves beyond engineering and becomes responsibility. A durable paver isn't just about long service. It's about reducing the frequency of replacement. Lowering the lifetime material throughput. Preventing carbon re-emission through avoidable manufacturing cycles. And giving cities, contractors, and designers a surface that can carry more than traffic—it can carry time.

## Final Thought

*Durability isn't a checkbox. It's a contract with the future.*

In the end, the question of durability isn't whether a paver is "strong enough." It's whether it was designed accordingly to last long enough to justify its environmental and economic footprint. It's whether the surface behaves like a system, not just a set of specs. It's whether the product can tolerate not only pressure and weather, but also the minor site negligence, imperfect installation, and unpredictable use. Because real-world durability doesn't reward perfection. It rewards resilience.

As we move forward in this chapter, we'll shift focus from the idea of durability to its mechanisms. We will focus on how stress, movement, traffic, and time wear a paver down, and what we can do to slow that process intelligently, not just expensively.

*"The most sustainable paver isn't the one with the highest test score. It's the one that's still doing its job—quietly, reliably, and without apology—long after the site was handed over."*

## 5.2 The Mechanics of Wear: Stress, Load, and Movement

*What breaks a paver first is rarely sudden—it's slow, invisible repetition.*

If durability is the outcome, wear is the process. And wear rarely announces itself. It doesn't start with a crack or a visible collapse. It begins in silence; in the subtle shifting of a joint, the settling of a corner, the grinding of dust beneath a tyre. It's the cumulative effect of pressure applied not once, but a thousand times. And it's precisely that repetition—more than any dramatic overload that wears a paving system down.

Interlocking concrete pavers are engineered to carry load. But they don't carry it the way monolithic slabs do. In a rigid slab, loads are spread across a continuous surface. In a modular system, loads are distributed and transferred unit by unit, across jointed edges, bedding layers, and the subbase beneath. Each layer has its job. Each interface matters.

And when one of them underperforms, the surface records that failure slowly, with every passing wheel or step.

## Understanding Load in Motion

*Because stress isn't static—and neither is failure.*

In the safe abstraction of design documents, load is treated as a number: axle weights, distributed pressures, $kN/m^2$. Clean. Predictable. Static. But on the ground, load is anything but still.

It moves. It accelerates, turns, stops, pivots, and reverses. It acts not just downward, but sideways. It doesn't arrive once—it arrives again and again, in the exact same places, creating stress concentrations that no spreadsheet can fully capture.

Take a delivery truck executing a tight turn at the edge of a logistics bay. On paper, it may meet load capacity. In practice, that turning motion applies torsional force across multiple units, scraping jointing sand sideways, rocking edge pavers in and out of alignment, and transferring lateral energy into the restraint system. A load that seems harmless in static terms becomes a source of gradual structural decay.

Even foot traffic, when high in frequency and concentrated in predictable pathways, becomes a form of dynamic wear. Retail plazas, transit corridors, school campuses—these are not soft-use zones. They're constant pressure systems, where every footstep carries grit, moisture, and micro-abrasion. If surface hardness and abrasion resistance weren't calibrated for that use, the failure won't be catastrophic; it will be slow, uneven erosion.

And the consequences don't happen in isolation:
- Jointing loss reduces lateral interlock
- Voids beneath the bedding layer amplify point load stress
- Repeated rocking creates micro-settlements, leading to visible undulation
- Water follows those deformations, worsening instability and enabling freeze–thaw cycles or biological growth

The paver may not crack. The mix may still meet spec. But the system begins to unravel—not from one big impact, but from tiny, repeated ones the design failed to anticipate.

*"Most early failures aren't structural defects.
They're behavioural oversights."*

This is why understanding load in motion is non-negotiable. It's not just about how much weight a surface can carry; it is about how that weight moves, and where it insists on returning. Because if your design treats load as static, it invites failure that's dynamic, and that's a guaranteed outcome.

## Predictable Stress, Predictable Failure

*Because what wears first is rarely a surprise—it's where you forgot to prepare.*

In the chaos of a construction site or a bustling logistics yard, it's easy to assume that stress is random—that failure could happen anywhere. But spend time observing how people move, how wheels turn, how weight travels across a space, and a pattern emerges. Stress is not chaotic. It's choreographed.

Forklifts don't turn in new places every day. They follow tight arcs, over and over again. Pedestrians don't wander aimlessly, they follow the shortest path between entrances. Delivery carts bounce along the same strip of pavement, guided not by design but by habit. These repetitive movements produce concentrated stress zones, and those zones are predictable. Which means failure is predictable too.

And yet many paving designs treat the surface as if the wear and tear will be evenly distributed. They're not only just ignoring the choreography of use—they continue to design against it.

A durable surface doesn't spread stresses evenly. It channels the stress intelligently and fortifies where needed. That requires a shift in mindset: stop designing for averages. Start designing for certainties. Because you already know where the problem will start.

So, the detailing must respond accordingly:
- Corners and turning zones should be reinforced; not just in the base, but in the jointing material and edge treatment
- Joint fillers should be specified not for appearance, but for their ability to resist lateral displacement
- Interlocking patterns; especially 45° herringbone, should be prioritized in any zone of repeated multidirectional force
- Subbase design must resist not just compression, but rutting, particularly in areas with limited drainage or soft subgrades

When failures occur in these areas, they're rarely because the paver was too weak. It's because the system was detailed as if load were abstract—not as if it would move exactly where it always does.

*"Stress has a memory. And it returns to the same place every day."*


## Movement Within and Around the System
*What moves beneath matters just as much as what stands above.*

Pavements are often imagined as fixed, unmoving surfaces—always stable, always firm, and always final. But beneath even the most precisely laid paver system, *movement is always happening*. Some of it is expected. Some of it is gradual. All of it matters.

Thermal expansion causes materials to swell and contract. Moisture levels in the subgrade rise and fall. Minor settlements occur under long-term loading. Even microscopic vibrations from nearby roads or equipment induce a low-grade creep. These aren't flaws in the system; they're features of the real world. And they require a surface that's not just strong, but forgiving. That's where the true function of joints begins.

Jointing isn't an aesthetic filler. It's the flexible ligament in a modular skeleton. The sand or aggregate between units is a shock absorber, a movement buffer, and a force distributor. If joints are too narrow, they act like rigid connections—transmitting stress instead of absorbing it. If they're too wide or poorly filled, they become erosion zones, and quick to fail under water, wind, or wheel. The system doesn't collapse because the concrete gave way. It collapses because the space between the concrete was never designed to move.

The bedding layer is just as critical. It's not only about levelling. It's the interface between structure and subgrade—the layer that must compress just enough to absorb load, but not so much that it settles over time.

If it's inconsistent in depth, uneven in compaction, or improperly drained, it becomes a silent saboteur. Minor voids grow. Differential settlement appears. Cracks form—not from failure at the surface, but from imbalance beneath it.

In permeable systems, the challenge intensifies. The bedding must support load and drain water simultaneously, a balance not all materials can maintain. If the stone is too coarse, it won't compact well. If too fine, it clogs. The wrong choice here turns performance into promise unfulfilled.

*"Rigidity may look strong—but it breaks under pressure. Systems that last are the ones that yield with intelligence."*

That's why the best paving systems aren't just designed for weight. They're designed for impermanence—for movement that's slow, inevitable, and survivable. Because what truly undermines durability isn't just traffic. It's ignoring the quiet shifts that happen underneath, every day, with no announcement—until it's too late.

## Case Insight: When Strength Isn't Enough

A detailed study published in the *Proceedings of the Institution of Civil Engineers – Transport* (UK, 2007) analysed the long-term performance of concrete block pavements in commercial and port logistics facilities. It found that pavements using stack bond or grid layouts in turning zones showed significantly higher rates of joint separation and surface deformation compared to those installed in 45° herringbone patterns.

The research, which included pavement sections at Tilbury Docks and Heathrow logistics yards, demonstrated that herringbone layouts distributed multidirectional forces more effectively, maintaining interlock integrity under wheel torsion. In contrast, stacked patterns concentrated stress along linear joints, leading to rutting and lateral shift under repetitive turning movements from forklifts and cargo vehicles.

Edge failures were also more prevalent where surface mortar was used for restraint, particularly in areas with poor drainage. Moisture ingress led to mortar degradation, joint widening, and eventual edge destabilization; regardless of the paver's compressive strength.

The conclusion was clear: "Structural failures were not caused by unit strength deficiencies but by poor interaction between load transfer geometry, jointing systems, and restraint."

## Designing for Repetition, Not Just Resistance
*Because surfaces don't fail from pressure alone; they fail from repeated pressure.*

To engineer durability is to stop treating wear as an anomaly. Repetition isn't a flaw in the system—it *is* the system. Every paver surface that sees regular use is already in a cycle of gradual degradation. The question is not whether wear will occur, but whether the system has been designed to absorb it gracefully, or collapse under its predictability.

This means the design process must shift from theoretical capacity to practical choreography. It's not enough to know how much load a paver can take in the lab. You need to know *where* that load will land, *how often, at what angle,* and *under what conditions.*

Designers must ask harder, more site-aware questions:
- Where will vehicles slow down, stop, and turn?
- What part of the layout bears the brunt of repetitive motion?
- The stress path aligned with the pattern's ability to distribute load?
- Has the jointing material been chosen for movement control?
- Are edge zones reinforced for torque and torsion, or just aesthetically framed?
- Do the bedding and base reflect the actual stress map of the surface, or were they drawn from a standard detail used everywhere and validated nowhere?

Because the goal is not to build a surface that resists wear at all costs. The goal is to build one that channels it intelligently. That shifts it, softens it, redirects it, and when needed—absorbs it without complaint.

*"Durability isn't invincibility. It's foresight. It's knowing where the first scuff will appear and making sure it stays only a scuff."*

In that sense, the most enduring pavements are rarely the strongest. They're the best-informed. Designed not for performance in isolation, but for performance in motion.


## Final Thought

*Wear doesn't come as a surprise. It arrives exactly where we failed to listen to the surface.*

The forces that wear down a paver aren't mysterious. They follow patterns, routines, and routes as familiar as the users themselves. Forklifts don't improvise. Pedestrians don't guess. Rainwater doesn't hesitate. Load is never truly random—it's rehearsed.

And so, the failures that follow are rarely surprises. They are the result of ignoring what repetition tries to tell us. A joint that wasn't wide enough. A pattern that wasn't meant to turn. A base that wasn't built to flex. A designer who assumed stress was a single moment, not a daily ritual.

Durability, then, is not the ability to resist forever. It's the ability to adapt over time, and to distribute stress, accommodate motion, and prepare not just for use, but for wear. And if wear is inevitable, failure doesn't have to be.

## 5.3 Environmental Exposure: Weathering, and Water

*Because the surface isn't just used—it's lived on by the climate.*

Long after the last compactor leaves the site, and long before the first sign of structural failure, a paver surface begins a quieter, slower battle with the elements.

Rain doesn't wait for drainage to be perfect. Sun doesn't care how well the pigment was stabilized. Algae and moss don't ask for permission before taking root in the shaded edges of a poorly sloped courtyard. This is the slow decay—the kind that doesn't announce itself until the surface is already eroding, discolouring, shifting.

And yet, environmental exposure is often treated as secondary; as an aesthetic concern, a long-term maintenance issue, a post-installation problem. But in truth, weathering is one of the first and most persistent forms of structural degradation. It creeps into joints. It expands small cracks. It stains porous faces. It gradually breaks down the cohesion between surface, base, and subgrade.

The challenge is not just to survive the elements. It's to design for them by anticipating that:
- UV will bleach surfaces and weaken exposed polymers
- Moisture will enter wherever it's allowed to linger
- Freeze–thaw cycles will force expansion from within
- Algae will thrive wherever drainage and drying are poor

And when these conditions aren't met with technical foresight—when slope, permeability, mix design, or surface finish are underestimated; what follows is not only visual degradation, but loss of performance.

In this section, we unpack the true cost of environmental exposure. Not as a cosmetic nuisance, but as a systemic stressor; one that must be designed into every layer if a surface is expected to endure.

## Water: The Slowest Structural Saboteur

*Because it's not how hard it rains—it's what the surface does with the rain that matters.*

Water is the most underestimated force in pavement design. It arrives without drama—quiet, persistent, often invisible in its damage. But given time, it outlasts concrete, shifts base layers, breeds biological growth, and turns fine workmanship into disrepair. Not by force; but by seepage, pooling, and repetition.

In many failed paver installations, water wasn't considered a structural problem. It was viewed as a drainage detail and left to civil consultants or sloped toward the nearest curb. But in practice, when water is allowed to sit, seep, or cycle through freeze–thaw conditions, it becomes the primary driver of failure.

It begins with something small with slightly compacted joint sand that retains moisture. That moisture seeps downward, saturating the bedding layer. Then, during the next heavy storm, or in temperate zones, a freeze–thaw cycle causes that trapped water to expand. Subtle lifting follows. The next loading cycle applies stress on uneven footing. Edges move. Voids form. The system shifts. And then comes the algae, the trip hazard, the "maintenance request."

### 1.   Drainage Is Not Optional—It's Structural

Water will find its way in. The design question is what happens next.

Durable paving systems treat drainage not as an external condition, but as an integrated property of the pavement system itself. This means:

- Surface slopes of 1–2% to move water off quickly
- Permeable joints or materials that encourage infiltration rather than ponding
- Bedding layers with controlled permeability—not so fast that they wash out, not so slow that they trap water

- Filter layers or geotextiles that prevent fine migration and clogging
- Subbases engineered to hold and release stormwater in balance—not just absorb on day one, but continue functioning for decades

In permeable systems especially, the water must not only enter; it must also leave. The moment outflow is restricted, the system behaves less like a sponge and more like a basin. One good storm turns a sustainable surface into a saturation trap.

## 2.  The Lifecycle Impact of Poor Water Management

Beyond the physical wear, water mismanagement accelerates maintenance cycles. Pavers that might have lasted 20 years begin showing moss within two years. Joints that should remain intact for five years need repointing every monsoon. And base layers begin to lose cohesion—not from use, but from being constantly wet.

This doesn't just affect durability. It affects carbon.

Every early failure triggers replacement: new materials, new manufacturing, new transport, new installation. And all because a surface wasn't designed to breathe, to drain, to recover from rain.

*"It's not the rain that breaks the system.*
*It's the design that pretended it wouldn't."*

## Efflorescence: The Subtle Surface Disruption

It begins subtly with a white haze, a chalky bloom across the surface. And yet for many clients, efflorescence sparks disproportionate concern. It's often misunderstood as a defect, a failure, or a sign of moisture ingress. In truth, it's a naturally occurring, non-structural phenomenon that reflects the chemistry of concrete and its interaction with the environment.

Efflorescence occurs when water-soluble salts; the calcium hydroxide from cement hydration migrate to the surface through the pore network of the paver. When this moisture evaporates, the salts react with atmospheric $CO_2$, crystallizing into visible white deposits, often within the first few months after installation.

In darker pigmented pavers, the contrast is more pronounced, leading to aesthetic dissatisfaction even when there's no performance issue.

What Triggers It?
- Frequent wetting and drying cycles post-installation
- New concrete with high lime content
- Poor drainage or base saturation
- Insufficient curing or sealing that allows free moisture transport

It is most commonly observed during the first 6 to 12 months, particularly in climates with alternating wet and dry periods. But its presence isn't inherently negative; it's a surface-level consequence of deeper hydration and transport dynamics.

## 1. Performance Impact: None, Visually Misleading

Efflorescence does not compromise the compressive strength, interlock function, abrasion resistance, or structural stability of a paver. Its only impact is perceptual, particularly in architectural, residential, or decorative applications. But perception matters; especially when it leads to callbacks, dissatisfaction, or premature replacement of otherwise sound systems.

## 2. Controlling the Occurrence

While impossible to eliminate entirely, efflorescence can be minimized by:
- Specifying low-alkali cement or using SCMs (e.g., GGBS, fly ash)
- Ensuring good drainage and capillary breaks in the subbase
- Avoiding oversaturation during installation
- Considering breathable surface treatments that reduce evaporative transport

### 3. Managing and Cleaning

Efflorescence is temporary in most cases. Normal weathering, rain, and surface use often fade it over time. Where cleaning is needed:

- Use mild acid solutions (e.g., diluted phosphoric acid) with careful application
- Avoid aggressive brushing or undiluted chemicals that can damage pigment or surface finish

A well-informed contractor can set client expectations early: what may appear is temporary, and it doesn't signal a deeper problem.

**"Efflorescence doesn't weaken the surface—it tests the strength of communication. Treat it as chemistry, not pathology."**


## Heat and UV: The Invisible Degraders

*Because sometimes the most damaging force is the one you can't feel— until it's too late.*

Heat doesn't crack concrete like ice. It doesn't flood joints like rain. It doesn't announce itself with puddles or stains. Instead, it arrives every day, incrementally. Quietly. Relentlessly. And over time, it begins to change the surface from the inside out.

In hot, high-radiation climates; especially in tropical or arid zones, interlocking pavers are exposed not just to elevated ambient temperatures, but to direct surface heating well beyond air temperature. Surfaces can regularly exceed 60°C–70°C, especially in darker-coloured pavers with low albedo.

At that temperature, even stable binders begin to dry out. Pigments fade. Sealants harden and crack. Micro-expansion causes slow joint widening and alignment shifts. And then there's UV.

Ultraviolet radiation breaks down polymers, weakens pigments, and triggers colour shifts; especially in unprotected or poorly stabilised surfaces. This isn't just an aesthetic concern. Fading often precedes surface chalking, abrasion loss, and a change in water absorption characteristics, all of which contribute to more aggressive biological colonisation and surface softening.

## 1. Heat Is Not a Passive Condition—It's a Stressor

In exposed urban settings, especially in equatorial or arid zones, the surface temperature of pavers routinely exceeds 60°C, with darker units often climbing past 70°C. These are not just numbers, they are thresholds for material fatigue:

- Cementitious binders begin to dry prematurely, weakening near-surface cohesion
- Bitumen-based or polymeric joint stabilisers soften, then harden irreversibly
- Expansion pressure builds between adjacent units, testing joint flexibility and edge restraint

And in poorly ventilated pedestrian corridors or high-density hardscapes, this radiant heat doesn't only damage the pavers, it also contributes to urban heat island effects, further straining surrounding infrastructure.

What looks like minor warping or pattern misalignment is often the accumulation of thermal cycles: daily expansion, nightly contraction, repeated again and again, until the surface memory slips.

## 2. UV: The Slow Chemical Undoing

Unlike heat, UV radiation doesn't cause expansion—it causes breakdown. Specifically, it:

- Degrades organic pigments, causing fading and discolouration
- Breaks down polymer chains in sealers, coatings, and some jointing materials

- Increases surface porosity over time, enabling moisture ingress and algae colonisation
- Alters tactile surface quality, leading to chalking, dusting, and surface weakening

These aren't hypothetical effects. In a 2020 study by the National Ready Mixed Concrete Association (NRMCA), paver samples exposed to continuous UV aging under ASTM G154 protocols showed up to 25% strength loss at the surface layer over 12 months; especially when untreated or sealed with non-UV-stabilised coatings.

## 3. Design and Specification for Solar Resilience

There is no one-size-fits-all answer to UV and heat durability, but there are design principles that mitigate their long-term effects:

- High-SRI aggregates and pigments lower surface temperatures, extending thermal lifespan and reducing pigment fade
- Use of iron oxide pigments (which are UV-stable) over organic dyes, especially for warm and dark tones
- Selection of jointing sands or aggregates that resist binding, dusting, or shrinking under high heat
- Expansion joints in large or high-exposure areas, especially in wide plaza or corridor layouts
- Surface treatments that are vapour-permeable, UV-resistant, and tested for climate compatibility

And just as importantly, installation practices must respect heat:
- Never lay pavers during peak heat hours if thermal bowing or premature curing is a concern
- Allow materials to acclimatise on-site before laying—particularly units stored in direct sun
- Moist cure sealed or pigmented units where appropriate to slow surface drying and protect finishes

**4.  Durable Surface Doesn't Ignore the Sky**

Too often, we design pavers to resist what's beneath them—load, subgrade, erosion. But real performance comes from designing for what's above them. For the sun, the heat, and the slow unrelenting chemistry of weathering. Because no matter how good the mix or strong the interlock, it is the surface—the skin of the product—that takes the first hit. And the next. And every day after.


## Biological Growth and Environmental Staining

*Because a surface that can't defend itself ends up cleaning itself—at your cost.*

No one includes algae in the design drawings. You won't find moss on a paving submittal, or mildew in a durability table. But after the ribbon is cut and the surface meets its first rainy season, it shows up anyway. In shaded corners. In high-humidity courtyards. In the neglected side path that no one sloped properly. And once it appears, it doesn't go away quietly.

Algae is not a cosmetic inconvenience; it's a systemic symptom. It tells you where water didn't drain, where the surface stayed damp too long, where air couldn't circulate, and where a material was chosen for appearance rather than resistance. And by the time someone notices, the damage is already layered: slipperiness, staining, softening of joint zones, and accelerated surface wear from aggressive cleaning.

### 1. Growth Follows Neglect, Not Just Nature

Biological growth like algae, mold, moss is not just some random occurrence. It thrives in places where three things converge:

- **Moisture that lingers**, either on the surface or within joints
- **Low UV exposure**, such as in shaded courtyards or narrow alleys
- **Surface porosity**, giving spores and organic dust something to cling to

Where these conditions persist, biological colonization is not a possibility—it's *inevitable*.

Walk any campus, public plaza, or shopping promenade in a tropical climate and you'll see the pattern. It's never the middle of the walkway that turns green. It's the spot by the planter. The strip near the downpipe. The alley behind the building where no one designed for sun or airflow.

That's the thing about organic growth: it tells the truth about your detailing.

*A surface that grows algae isn't failing—it's just doing exactly what it was allowed to do.*

### 2. Material Science vs. Microbial Persistence

Some materials handle exposure better than others. Low-porosity units, engineered to shed water quickly and resist staining, can hold their own for years. Others—particularly those with highly textured, absorbent finishes—become perfect substrates for growth, especially when untreated.

Modern technologies offer ways to slow the spread:
- **Hydrophobic surface treatments** reduce wetting time and discourage microbial anchoring
- **Micro-structured textures** mimic anti-fouling properties found in nature, such as lotus leaves

- **pH-modified cement blends** create inhospitable surfaces for algae to colonize
- **Photocatalytic additives**, often titanium dioxide-based, actively break down organic material when exposed to UV (effective in sunnier climates)

But even the best formulations can't overcome poor detailing. Stagnant water always wins.

## 3. The Hidden Cost of Cleaning

Once growth begins, cleaning becomes a cycle—one that adds environmental, financial, and structural stress:

- **Pressure washing** strips joint sand and surface texture
- **Chemical cleaners** introduce runoff concerns, particularly near storm drains or softscape zones
- **Repeated cycles** shorten surface lifespan and increase maintenance cost
- **Slip risks** increase liability, especially in schools, hospitals, and high-footfall areas

This is where biological exposure becomes more than a maintenance concern—it becomes a *durability multiplier*. Not because the material failed, but because the system allowed conditions that accelerated wear.

*"A surface that requires monthly cleaning is not a surface built to last, It's a surface outsourcing its resistance to your maintenance team."*

## 4. Designing for What Will Grow

You cannot design outdoor surfaces as if they'll stay clean. You design for where the dirt will go. Where the water will pause. Where the spores will land. And most importantly, where the surface will dry or won't.

Good design means:
- Sloping everything—not just the middle of the walkway, but the edges too
- Avoiding layouts that create airless, sunless corners
- Choosing jointing systems that can shed moisture, not store it
- Using textures and finishes that resist not only traction loss, but biological adhesion
- Being honest about who will clean this, how often, and with what equipment

Because algae aren't a cleaning issue. It's a surface story, and every growth tells you where someone forgot to listen.


## Final Thought

*Surfaces don't fail because of weather. They fail because weather was treated like an afterthought.*

Rain is not a surprise. Heat is not a rare event. UV is not invisible in the data. And algae don't appear without warning. These forces are natural, expected, and are repetitive. They're everyday realities. And yet, they're too often relegated to the margins of the design brief, the "post-installation concerns," the maintenance chapter no one reads.

But this section makes clear: environmental exposure is not cosmetic—*it is structural*. It wears, weakens, stains, and saturates. It makes good materials look neglected, and exposes detailing that wasn't designed to endure.

The most resilient surfaces aren't those that avoid the elements. They are those that engage with them intelligently. Surfaces that expect to get wet, and know how to dry. That anticipate UV, and don't fade. That resist growth not by sealing out nature, but by shaping its interaction with the built world.

*"Durability isn't built for the lab.*
*It's built for the rain, the heat, and the shade you forgot to calculate."*

Because if we want surfaces to last, we must stop treating the climate as something we design around, and start designing for it.

## 5.4 Failure Modes – And What They Teach
*Because every crack has a backstory—and it's always worth reading.*

Durability is most often defined by what survives. But it's best understood by what fails.

Cracks, corner breaks, sunken edges, faded zones, rattling pavers; these aren't just annoyances or quality lapses. They're signals. Each one is a visible artifact of an invisible oversight. A reminder that somewhere in the chain; from design to detailing to installation to use, something was underestimated, ignored, or idealized.

Yet failure is not the enemy. In fact, it's the most honest feedback loop we have. A failed surface doesn't lie. It doesn't flatter. It tells you exactly where your assumptions fell short.

That's why this section doesn't just diagnose failure modes. It listens to them. Because if every cracked paver is treated as a nuisance to replace rather than a lesson to be analysed, we lose the opportunity to build smarter, longer-lasting systems.

Over the next few pages, we'll walk through the most common (and revealing) failure types—not just to understand what went wrong, but to uncover what could have been done differently:

- Not just that the edge cracked, but why it cracked *there*
- Not just that the joint failed, but what it was never designed to handle
- Not just that the surface faded, but how exposure outpaced specification

***"A failed surface is not a lost investment,***
***It's an unpaid bill from a decision made upstream."***

## Cracking and Corner Breakage

*Because when concrete fractures, it rarely breaks on its own.*

Cracking is often seen as the most obvious failure in concrete surfaces. It's visible. It's measurable. It can be photographed, annotated, and easily blamed. But in the world of interlocking pavers, cracks rarely originate from within the paver itself.

They begin in the system. In movement that wasn't restrained. In bedding that wasn't level. In joints that transferred too much stress. In expansion that had nowhere to go. When the system is out of balance, the unit pays the price.

Corner breakage, in particular, is one of the most revealing failures in modular pavements. It typically occurs at high-stress points; where loads transition, turn, or concentrate at edges. And the moment a corner snaps, the local geometry of load distribution changes, leading to accelerated wear on adjacent units. A single broken unit is rarely alone for long.

## 1.  Diagnosing the Real Causes

While a cracked paver may appear to be a material defect, the root cause almost always lies elsewhere:

- **Poor bedding support** results in point loading, especially at the paver's edge or corner
- **Loss of jointing sand** increases lateral movement and reduces surface stability
- **Subgrade settlement** introduces uneven support, placing shear stress on thin points
- **Overloading** beyond the design capacity concentrates force on vulnerable zones
- **Thermal expansion** without sufficient expansion joints causes compression cracking at unit boundaries

In freeze–thaw regions, moisture ingress into microcracks and further expands damage. In tropical zones, repetitive thermal cycling can lead to fatigue; especially in lighter, thinner units or poorly restrained edge areas.

*"The paver doesn't crack because it's weak. It cracks because it's alone where it was supposed to be supported."*

## 2.  Prevention Is in the Pattern, Not the Patch

Cracks don't ask for patching, they ask for rethinking. Durable paver systems resist breakage not by using thicker units alone, but by:
- Ensuring uniform bedding compaction across the entire plane
- Reinforcing edge restraints to resist uplift, rotation, and lateral creep
- Selecting interlocking patterns that distribute stress diagonally (herringbone over stack)
- Avoiding long, unbroken runs of pavers without transitions or expansion joints
- Planning for thermal and moisture movement—not just structural load

And when breakage does occur, don't just replace the unit. Replace the conditions that allowed it to fail. Because without systemic correction, the next one is already on the clock.

*"A cracked corner is a structural whisper: "You missed something upstream." Listen before the next one speaks louder."*


## Joint Failure and Edge Spreading

*Because when the connections fail, the system forgets how to hold itself together.*

In any interlocking pavement, the joint is not a gap. It is the mechanism. It's the reason why modular paving works; the joints absorb stress, transfer loads, buffer movement, and give the surface room to flex without fracturing. But when those joints fail; whether by erosion, compaction, or contamination, the entire surface begins to behave differently. **The pavers don't crack immediately. They drift. They rise. They settle. And eventually, they separate.**

Edge spreading is the inevitable consequence. Without joint integrity, interlock is lost. Without interlock, restraint becomes localized. And without system-wide restraint, the edge becomes the first casualty.

*"Modular paving is only as strong as the space between the modules."*


### 1.  What Causes Joint Failure?

Joint failure rarely comes from a single cause. More often, it's the **slow collapse of joint function** over time:

- **Washed-out jointing sand**, especially in high-rainfall or poorly sloped areas
- **Insufficient compaction** during installation, leaving joints shallow and loose
- **Wrong aggregate gradation**, allowing fines to compact or displaced
- **Traffic vibration** that causes gradual pumping of joint material

- **Weed and root intrusion** in neglected installations, forcing pavers apart
- **Cleaning abuse** from overuse of high-pressure water or corrosive cleaners

The result is a progressive breakdown of the pavement's ability to behave as a unit. As interlock fades, individual units begin to shift under load. This shift is rarely dramatic at first; it begins with minor rotation, tiny vertical movement, or subtle gaps at the edges. But these changes accumulate and eventually, edge pavers start to "walk", joints widen, corners chip, load transfers poorly, and surface deformation becomes visible.

## 2. Designing to Preserve the Joint

If the paver is the structure, then the joint is the dialogue. It's where load is negotiated, movement absorbed, and the illusion of rigidity sustained. And like any dialogue, when the joint is weak, 'miscommunication' leads to disintegration.

Preserving joint performance isn't about overengineering; it's about understanding that durability lives in the spaces we tend to overlook. The strongest surface can fail if the material between its parts erodes, compacts, or disappears entirely. And once those voids form, every load applied begins to act individually on each unit, rather than being absorbed collectively across the system.

> *"A surface doesn't collapse when its pavers weaken;*
> *It collapses when the space between them loses meaning."*

Durable joint design begins with intentional material choices:
- Hard, angular aggregates that lock into place under compaction
- Gradations engineered to resist movement under traffic and vibration
- Hydrophobic or stabilised sands where washout or biological growth is likely
- Avoidance of soft or rounded grains that behave more like filler than framework

But material is only one layer. Joint geometry, system slope, drainage exit points, and traffic type all influence how long that joint holds. If the slope directs runoff directly across exposed joints, no aggregate will hold indefinitely. If vibration from turning vehicles is constant, even the best jointing will need re-tensioning. If the installer over-sweeps or under-compacts, the space becomes a failure point waiting to happen.

Preservation, then, is not a product spec—it's a system condition. And most critically: edges matter, not just as borders, but as boundaries of containment. When joints begin to lose cohesion, the first units to drift are the ones without enough resistance.

### 3.  Maintenance as Design, Not Reaction

Too often, joints are treated as finished once they're swept. But in reality, post-construction joint stability is an active phase of durability. Surfaces need time to settle, and joints often need re-topping or light re-compaction after the first few weeks of service; especially in permeable systems or on uneven subgrades.

And in long-term planning, maintenance shouldn't be reactive. It should be baked into the design intent:

- What is the expected joint loss over five years?
- Who will inspect it?
- What's the reapplication process?
- Can it be done without major disruption?

Because preserving the joint isn't just a technical exercise. It's a philosophical one. It's choosing to value what connects, not just what carries.

*"Design for what ties the surface together,*
*and it will hold, long after the load has passed."*

## Surface Abrasion and Colour Fading

*Because a surface that loses its identity often loses its value next.*

Not all failure is structural. Some of the most telling signs that a surface has given up show not in what breaks but in what fades. A surface that once looked crisp, vibrant, and tight in texture slowly becomes dull, dusty, inconsistent. It begins to feel different underfoot. It loses the finish that made it seem intentional.

Surface abrasion and colour fading don't often trigger immediate repair orders, but they quietly erode the perceived quality of a space. In retail environments, civic plazas, campuses, and commercial walkways, this matters more than we admit. When a surface begins to fade, so does the trust in the space it's part of.

This isn't just an aesthetic issue. It's a performance one.

## 1. How Abrasion Happens: A Slow Disassembly

Abrasion isn't about impact. It's about repetitions like the soft, slow grind of footfall, wheeled traffic, wind-blown grit, and periodic cleaning. Surfaces don't wear down because they're used violently. They wear down because they're used constantly.

Here's how it begins:
- Slight micro-wear along aggregate peaks
- Fine particles loosened by shoes, sand, and moisture
- Increased porosity on the surface layer
- Faster water retention, leading to biological growth
- Even more cleaning—creates more abrasion

In time, what remains is a paver that's still structurally sound, but visually and texturally degraded. In design terms, it has outlived its impression, even if not its strength.

## 2.  Why Colour Fades, and Why That Matters

Fading is often dismissed as inevitable. And yes—*all* surfaces weather. But not all colour loss is created equal.

Pigments that are not UV-stabilised begin to degrade under solar exposure. The surface begins to chalk, and the binder matrix near the top weakens. In pavers with integral colour, this process can reveal raw, unpigmented aggregates, especially if the mix wasn't properly blended. In coated units, the loss of surface sealer accelerates both pigment fade and textural disintegration.

This doesn't just look tired—it changes how water interacts with the surface, how traction behaves when wet, and how quickly biological staining occurs.

## 3.  Designing for Appearance that Endures

If a paver is meant to perform for 20+ years, its appearance must hold its own for at least half that time. A surface that fades or wears within the first five years undermines every sustainability claim attached to it because even if it doesn't crack, it will be replaced for how it looks.

Durability in appearance is not vanity; it's public trust in material integrity.

To preserve that:
- Choose UV-resistant pigments, not just appealing shades
- Ensure colour is integrally mixed, not superficially applied
- Specify densified wearing layers or factory-applied surface hardeners
- Avoid over-polished or overly porous textures in high-use zones
- Use vapour-permeable sealers only where needed, and plan for reapplication
- Don't ignore surface SRI or light reflectance values—these fade too

*"In high-exposure zones, a fading surface isn't wearing out.*
*It's being worn down by assumptions that the sun wouldn't matter."*

## Subsurface Deformation and Settlement

*Because what you don't see moving is often what moves everything else.*

The surface may carry the traffic, but it's the unseen layers beneath that carry the responsibility. And when those layers shift, settle, or fail to hold form, even the best-designed surface will follow them into misalignment.

Subsurface deformation isn't dramatic, at least not at first. It doesn't crack loudly or break visibly. Instead, it appears as subtle undulations. A low spot that holds water. A rocking paver where once there was none. Edges that tilt just enough to catch the foot. These aren't cosmetic flaws; they're symptoms of deeper structural drift.

And once that drift begins, the pavement is no longer a surface; it's a slow-motion collapse in disguise.

### 1. What Causes Subsurface Instability?

Subsurface deformation is most commonly driven by three interlocking failures:

1. **Inadequate compaction**—either of the subgrade or subbase, leaving trapped air pockets or loose zones
2. **Variable moisture content**—soils that swell and shrink unevenly, especially in clay-heavy or poorly drained locations
3. **Dynamic load concentration**—areas with repetitive point loading (like delivery zones or bus stops) that exceed the long-term bearing capacity of the base

Other contributing factors include:

- Use of non-graded or poorly drained aggregates
- Absence of geotextile separation, leading to subgrade contamination of base layers
- Edge loss or perimeter leakage that allows pavers to "walk" under cyclic stress

In permeable installations, deformation is even more sensitive. Void-retaining layers, if not properly compacted or protected from fines intrusion, can collapse under hydraulic pressure or saturation cycles. Settlement then manifests not just as visual sag, but as hydrological failure: ponding, erosion, and water bypass.

*"A paver system doesn't deform because it was weak.*
*It deforms because what held it up stopped holding."*

## 2. Detecting the Drift Before It's Too Late

Subsurface deformation is easiest to detect when it's already problematic. But some early indicators include:

- **Persistent surface moisture** in areas without visible slope errors
- **Differential joint gaps** along previously aligned zones
- **Early rattling or rocking** in isolated units
- **Efflorescence streaks or moisture marks** along edge courses

These aren't defects in the surface—they're signals from below.

## 3. Designing for Stability Below the Surface

True durability starts deep. If the base fails, the surface will only mask the problem, until it no longer can.

Designing to prevent subsurface deformation means:
- **Compaction is "king"**—validated with field density testing, not just eyeballing
- Use of **graded, angular aggregates** that interlock and drain—not rounded or gap-graded fill
- Inclusion of **separation geotextiles** between subgrade and base to prevent upward migration of fines
- Ensuring base layers are **deep enough and uniform**—not tapered or thinned to "match levels"
- Designing for **traffic type, frequency, and long-term loading**—not just design vehicle specs

And most importantly, recognizing that settlement is slow sabotage. It happens quietly, but costs loudly.

*"If your surface feels off-balance, the imbalance started far beneath it. Fixing the top won't fix the truth underneath."*


## Water Retention and Drainage Failures
*Because what stays on the surface rarely stays harmless.*

Water doesn't need to arrive in torrents to leave its mark. It only needs to stay. A thin film left behind after rain, a subtle depression that doesn't drain, a patch that never quite dries—these are not benign. They are the first moves in a slow but deliberate sequence of deterioration.

The danger isn't the water itself—it's what it invites. Lingering moisture changes the chemistry of the surface. It darkens tones. It softens joints. It welcomes spores and seeds. Over time, what should have been a resilient, breathable surface becomes a breeding ground: for algae, for erosion, for settlement.

What looks like a puddle is often the surface asking for help. And yet, this kind of failure is among the most overlooked. Not because it's rare, but because it's so predictable. Of all the ways an interlocking paver system can fall short, retaining water is one of the most preventable one.

## 1.  What Keeps Water from Leaving?
Water retention doesn't happen because of one big mistake. It happens because of a dozen small oversights that might seem harmless on their own but together create a system that quietly refuses to drain. Sometimes it begins with a slope that looks clean on paper but flattens out in execution; designed more for aesthetics than for flow. Or it creeps in through uneven bedding: a slight dip here, an over-compacted patch there, enough to let water settle rather than move.

In permeable systems, the problem often hides deeper. Joints begin to clog; not because the material was wrong, but because no one planned for the fines, leaves, or sediment that would inevitably arrive. Base layers, once open-graded and breathable, slowly seal themselves off. What once infiltrated now stagnates.

Edge restraint errors make things worse. Misaligned perimeters tilt surfaces inward, encouraging pooling instead of shedding. Water obeys the new geometry. It stays where it shouldn't.

And in hybrid systems—where permeable and impermeable zones meet—there's often no plan for what happens at the boundary. Overflow paths are missing. Underdrains are forgotten. The surface, designed in parts, never quite becomes a whole.

*"These aren't material failures. They're failures of anticipation of assuming water would behave, without being told how to."*

## 2. Reading the Surface: Early Signs of Drainage Failure

Drainage problems rarely arrive with a headline. They announce themselves in whispers, that something deeper isn't working as it should be.

You might notice a patch that stays darker long after the rain has stopped. Not quite a puddle, but never fully dry. Or the slow creep of algae in corners that were meant to stay clean. On closer inspection, the joints may feel damp to the touch, the sand eroded or missing altogether. Sometimes, the paver moves slightly underfoot—an unfamiliar softness where there used to be certainty.

White bands may appear along the surface; efflorescence rising up, drawn by trapped moisture struggling to escape. Along the edges, moss and algae quietly takes hold; particularly where shade, trees, or poor ventilation allow moisture to settle longer than it should.

Individually, these signs might be dismissed as maintenance issues. Collectively, they tell a different story: that water is no longer moving the way it was supposed to.

*"These aren't just blemishes. They're surface expressions of a system beginning to slip out of balance."*

## 3.  Designing Water Out, Not Just Letting It In

Too often, surface design treats water as something to tolerate. As long as it doesn't flood, it is assumed that the system is working. But true durability doesn't come from passive resistance. It comes from active movement; from creating a surface that doesn't just accept water, but manages it, channels it, and finally, lets it go.

That begins with slope; not just in theory, but in execution. A surface with even a 1% fall can be enough to keep water moving. But that slope must be everywhere, not just in the drawings; it has to be on every stretch, every courtyard, every transition.

In permeable systems, it's not enough for water to enter. It must have somewhere to go. Without well-positioned outlets, underdrains, or flow paths through the base layers, infiltration becomes storage. And storage becomes saturation. The base that was meant to breathe becomes a sump.

Jointing material plays its part too. It must remain open long enough to allow flow, but firm enough to resist migration. Too loose, and fines wash out. Too tight, and clogging begins. Balance is essential, and it begins with aggregate gradation and not just guesswork.

At the system's base, geotextile separators and filtration layers are what keep order. Without them, fines rise, voids collapse, and permeability dies one rain at a time. The system may still look intact, but it's already lost its function.

And then comes the overlooked factor: maintenance. No surface lasts forever without care. But the best designs build that care into their logic by incorporating accessible joints, removable grates, planned access points for vacuum extraction or manual cleaning. If no one can reach the clog, no one can prevent the pond.

So every design should ask, not "can this drain?", but "when it clogs, how will we fix it?"

Because in every sustainable pavement system, water is not the enemy—Stagnation is. And the best way to build resilience is to give water a clear, deliberate exit.

**Final Thought**
*Surfaces don't fail all at once. They fail in stages—and they always leave clues.*

In every project, the pavers arrive perfect. The joints are clean. The layout is crisp. The finish looks ready to endure. But performance isn't about how things begin; it's about how they respond when the real conditions arrive.

The failure modes that we have examined so far; cracking, joint erosion, fading, settlement, drainage breakdown aren't just unfortunate accidents. They are predictable outcomes when systems are treated as collections of parts, rather than as integrated responses to load, weather, water, time, and use.

Each failure we studied began with a missed conversation:
- About what would move and where
- About how traffic would behave
- About how water would be shed or not
- About what would shift beneath the surface
- About what the system would become after a few thousand cycles of pressure

If we listen carefully, each failure doesn't just show us what went wrong. It teaches us how to design differently next time with more patience, more anticipation, and more respect for the quiet stresses that live in every paved surface.

## 5.5. Design for Durability: Prevention by Planning
*Because what doesn't fail is usually what was never left to chance.*

Durability isn't created on-site. It's created upstream; at the desk, on the drawing board, in the specifications, and within the questions asked long before any paver touches soil.

By the time a surface fails, the damage isn't just physical—it's procedural. Somewhere along the chain, a detail was rushed, a condition was assumed, or a load was idealised. That's why prevention is never just about stronger materials or thicker layers. It's about better thinking, earlier.

This section is not about adding safety margins. It's about shifting the entire approach—from reaction to anticipation. Because durable paving is not about designing to resist failure; it's about designing with enough clarity that failure never gets a foothold.

Durability lives in:
- The slope you confirm, not just the one you draw
- The joint you specify for movement, not just alignment
- The base you compact with testing, not trust
- The drainage you verify with flow paths, not hope
- The wear zones you reinforce because people will turn

What fails is what was assumed. What survives is what was planned.

*"Prevention isn't about overdesign. It's about respecting reality early enough to respond to it with intent."*

## Detailing for Stress and Movement
*Because stress is not a surprise—and movement is not a failure.*

Every paved surface will be stressed. That much is guaranteed. What matters is whether it was designed to handle stress gracefully, or to resist it blindly.

Too often, stress is treated as an exception—something to address only in high-load zones or critical intersections. But stress exists everywhere. It lives in the rhythm of daily foot traffic, in the turning arc of a forklift, in the shifting of a clayey subgrade after rain, in the expansion of hot concrete under noon sun. And it always moves.

The misconception is that movement equals failure. But in high-performing interlocking systems, movement is accommodated—not denied. Joints are expected to flex. Bases are expected to absorb.

Patterns are expected to distribute. The system doesn't hold still; it holds together.

Durability begins in the detailing; not just in specifying materials, but in mapping how force will flow through the system. That means:
- Recognizing where load will concentrate—not just average out
- Anticipating how thermal expansion will push and pull across large runs
- Accepting that joints must be both tight enough to stabilize, and wide enough to shift
- Designing patterns that encourage interlock, not just symmetry

Take the common oversight of using grid layouts in turning zones. While it may look neat on plan, it fails in reality, where load moves diagonally. A 45° herringbone pattern in those areas could have offered multidirectional stress distribution. But without that decision in the detailing phase, the system begins its life at a disadvantage.

Even more critical is the treatment of edge restraints. When they are underspecified or treated as an afterthought, the whole paving field begins to float. Lateral creep accelerates. Joints widen. And the most stressed units often at the periphery begins to chip, tilt, or lift. Not because the pavers were weak, but because the edges didn't hold the pressure in.

And then there are expansion joints. While common in concrete slabs, they are often forgotten in modular systems under the false belief that joints between units are enough. But in large, exposed areas subject to heat or swelling soils, controlled points of release are critical. Without them, movement expresses itself through buckling, cracking, or dislodgement.

This level of detailing takes time. It requires not just referencing standard drawings but interrogating them. Questioning every radius, every slope break, every transition between material types. Because that's where stress gathers. And if it wasn't given an answer during design, it will find one during use. And that answer usually looks like failure.


## Material Pairing and Performance Matching
*Because even the best material will fail in the wrong context.*

There's no such thing as a universally durable paver. There is only a paver that is well matched—to its load, to its environment, to its installation method, and to what it will be asked to endure over time.

Too often, the conversation around durability begins and ends with compressive strength. But strength, on its own, is a static metric. It doesn't tell you how a surface will handle frictional wear, or how it will react to constant moisture, or whether the pigment will hold after five years of UV exposure. It doesn't tell you if the jointing sand will migrate. Or if the subbase aggregate is stable enough to keep everything above it in place.

**"Durability is never about a single component.
It's about compatibility across the system."**

A high-strength paver placed on an unstable base will still sink. A highly permeable system using fine, clog-prone bedding will still fail. A UV-stable surface that sits above a saturated subgrade will still buckle, tilt, or displace. The problem isn't the product—it's the pairing.

Take, for instance, a pedestrian plaza in a tropical climate. The selected paver may have excellent skid resistance and surface finish, but without algae-resistant properties or a hydrophobic additive, its appearance will degrade quickly; not because it was a poor choice, but because it wasn't matched to the bio-environmental conditions.

Or consider a high-SRI, light-coloured unit specified for heat island mitigation in an urban streetscape. If paired with a jointing material prone to erosion or if cleaning methods involve high-pressure washing, the surface may lose both colour and stability within the first few years. The technical specification may have looked sound. But the interplay between materials, use, and maintenance was overlooked. This is where ***performance matching*** becomes critical.

Designing for durability means not just asking how each material performs on its own, but how it behaves in relationship:

- Will the jointing sand stay in place under expected water flow rates?
- Will the pigment system resist UV degradation at the project's latitude?
- Will the binder react chemically with airborne pollutants or irrigation overspray?
- Will the aggregate gradation support long-term permeability without collapse?
- Will the sealant or coating interfere with drainage or vapor diffusion?

These aren't secondary questions. They are **the first filter** through which every design decision should pass.

True durability is born when the materials not only meet spec sheets but meet each other with understanding. The right combination becomes self-supporting. It flexes together, ages together, and survives together. The wrong one? It undermines itself from the moment installation begins.

## Designing for Maintenance, Not Just Installation
*Because what stays durable is what was meant to be cared for.*

**Installation is a day...Maintenance is a decade.**

Yet in most paving designs, maintenance barely earns a line item. It's often assumed that if the surface is built "to spec," it will stay that way; regardless of how it's used, weathered, or cleaned. But the truth is, a surface that isn't designed with maintenance in mind is already on its way to failure.

Durability doesn't end when the last compactor rolls off-site. In reality that's where it begins. And what determines how long that durability lasts aren't just material strength or installation precision; it's how easily the system can be inspected, cleaned, adjusted, and preserved without disruption or deterioration.

A high-performance paver system should be designed for re-entry. For people to return; not just to walk on it, but to maintain it.

That means thinking ahead:
- Can jointing sand be topped up without damaging surrounding units?
- Is the system compatible with vacuum sweepers or water-based cleaning tools?
- Are there accessible points for inspecting and flushing underdrains in permeable designs?
- Will the edge restraints resist degradation from repeated equipment use or mechanical vibration?
- Are sealers or surface treatments re-applicable without full-area shutdown?

None of these are post-handover questions. They are design-phase decisions, and they should be asked as early as the first layout sketch.

Consider jointing aggregates: in areas prone to water flow, choosing a polymer-modified or stabilised joint filler might cost more upfront. But if it means avoiding biannual repointing or sand washout, it's a durability investment disguised as a maintenance decision.

Take permeable pavement systems; without pre-planned access points for sediment removal or underdrain flushing, the first clog becomes a major intervention, turning a routine inspection into a full restoration. And in surfaces that blend permeable and impermeable zones, forgetting to plan for sweeping patterns or water routing turns maintenance into a compromise: clean one area at the expense of another.

*"If your surface isn't maintainable, it isn't sustainable.*
*It's just temporarily clean."*

Designing for maintenance means extending the lifecycle; not just of materials, but of decision-making. It means building in flexibility:
- surfaces that can be lifted and reinstalled without damage,
- edges that can be reset without excavation
- joints that can be refreshed without risk of over-compaction.

It also means designing with the user in mind. Not just the end-user, but the team who will return every six months with tools in hand and a schedule to keep. The real test of design isn't how well it installs, but how easily it survives the people assigned to look after it. Because they are not maintaining the material. They are maintaining your decisions.

## Anticipating Use, Abuse, and Change Over Time

*Because surfaces are rarely used the way we imagine—and they never stay in their original context for long.*

Design drawings are clean. They show pavers aligned perfectly, patterns uninterrupted, traffic loads evenly distributed. But real life is uneven. People cut corners. Forklifts drift wide. Trash bins leak. Trees grow. Deliveries arrive late, and no one uses the turning radius as planned.

Surfaces are not used as how it was designed; they're always used as lived in. And that reality is often the difference between a surface that lasts, and one that doesn't make it past its fifth year without a rework.

Durability isn't just about stress resistance or material compatibility. It's about understanding what time will do to a place, and how the system will respond when the context changes.

That means thinking beyond the first layer of use:
- Will food stalls show up during weekends?
- Will landscaping overflow during rainy seasons, dumping mulch and silt across walkways?
- Will loading bays be expanded in the future, placing new pressure on once-pedestrian surfaces?
- Will shading patterns change as trees grow, increasing algae risk in new places?
- Will maintenance practices shift with budget cycles, reducing cleaning frequency or changing equipment?

These aren't hypotheticals. They're the lived realities of hardscapes in the wild, and every one of them carries a durability consequence if the design doesn't build in room for adaptation.

Designing for long-term performance requires us to accept that abuse is part of use. A plaza edge will get driven over. A corner paver will be pried up. A high-footfall path will wear faster than surrounding areas. The question isn't whether these things happen; it's whether the system was detailed to recover from them.

Modularity is one answer. Reinforced edge restraints are another. Staggered patterns with high interlock offer resilience under rotational stress. Using the same paver format across multiple zones allows for localized repair without visual disruption. And incorporating zones of flexible material like resin-bound aggregates or grouted concrete strips in high-abuse areas can absorb shocks the pavers weren't meant to take.

*"Design doesn't end when the plan is printed.*
*It continues in the realities we didn't fully predict and prepare."*

Time will challenge the surface. So will people. So will neglect. Durability comes from giving the surface the tools to respond, not with resistance, but with resilience.

## Final Thought

*Prevention isn't a feature of the surface—it's a feature of the mindset behind it.*

What lasts isn't what we build the strongest. It's what we understand the best.

Durability, in its most complete form, isn't something you add to a paving system. It's something you embed through questions, foresight, and detail. It begins long before materials are selected. Long before installation begins. It starts when we stop designing for the ideal and begin designing for the inevitable.

When we anticipate wear, rather than react to it.
When we accept movement, rather than fear it.
When we plan for maintenance, rather than leave it to others.
When we treat changing use patterns not as threats, but as inputs.

The systems that endure are not overbuilt; they are over-considered. Every layer has a purpose. Every detail earns its place. Nothing is left to chance, because chance has already been accounted for.

> *"Good design installs a surface.*
> *Great design installs a future it's ready for."*

And that's what separates temporary success from long-term value: not how well the surface performs today, but how confidently it moves through the pressures of tomorrow.

## 5.6. Maintenance Cycles and Whole-Life Costing

*Because the true cost of a surface only reveals itself long after it's installed.*

At handover, every paved surface looks finished. But what happens next—*how it ages, how it's cleaned, how it responds to wear, and how often it needs attention*—is where the real cost begins to emerge. Not in upfront numbers, but in cycles.

Maintenance isn't the afterthought of durability—it's the test of it. And whole-life performance isn't measured in strength tests alone, but in how little a surface demands over time to keep functioning, looking, and performing as intended.

Too often, project budgets are structured around capital cost alone. Design decisions are made to reduce the bill of quantities, minimize installation time, or optimize immediate performance metrics.

But when paving systems begin to show wear—when joints loosen, surfaces fade, or algae takes hold; the cost to restore can be bigger compared to what it would have taken to prevent.

This is why maintenance must be planned—not assumed. And why whole-life costing needs to become a foundational part of sustainable hardscape strategy.

This section reframes durability not just as a technical property, but as an economic cycle:

- What interventions will be needed—and when?
- How often will joints require re-filling, surfaces need resealing, drains need clearing?
- What's the cost of a reinstallation after 8 years versus a thoughtful design that lasts 20?
- How do energy, water, labour, and downtime factor into true lifecycle cost?

Because sustainability doesn't stop at embodied carbon. It continues in how often the product must be touched, fixed, cleaned, or replaced, and what resources are consumed each time.

*"If a surface needs constant attention to look like it's working,*
*it isn't working—it's borrowing performance on credit."*

## Maintenance as a Measurable Performance Indicator

*Because if you can't measure how often it needs attention, you can't claim it performs.*

Durability has long been framed in terms of strength, modulus, or wear resistance. But in the real world, performance doesn't only show up in test results—it shows up on schedules, on calendars, and on invoices. A truly high-performing surface is one that stays out of the maintenance log.

It's easy to underestimate the role that maintenance plays in defining success. When we specify a paver with a high compressive strength or a pigment with long-term UV stability, we assume performance will follow. But that's only part of the equation. If the jointing fails after every monsoon, if algae build up in shaded walkways by the second rainy season, if patch replacements create visual inconsistencies; then performance has already dropped, no matter what the data sheet says.

This is where maintenance becomes a proxy for resilience. The less frequently a surface needs intervention, and the less invasive that intervention is when it comes, the more successful the system is—not just technically, but economically and environmentally.

Forward-thinking projects now treat maintenance frequency and intensity as key metrics in evaluating paving system options. That means tracking:

- How often jointing needs to be topped up
- Whether cleaning requires chemical treatment, pressure washing, or simple sweeping
- How long surface treatments like sealers remain effective under actual site conditions
- The time, skill, and tools required for minor repairs
- The lifecycle of edge restraints or base compaction before realignment is necessary

When these metrics are built into asset management plans and tied back to specific design decisions, they transform from burdens into tools. The client no longer just sees the paver as a static material. They see it as part of a system that either generates operational efficiency— or quietly drains it.

Just as energy efficiency is measured in kilowatt-hours saved, surface durability should be measured in interventions avoided, in man-hours not spent, in chemicals not used, in public complaints not received.

This is how paving becomes not just a product but a strategic asset. And how maintenance, when properly planned and tracked, becomes the most honest scorecard of design quality.


## Planning Intervention Cycles by Surface Type and Use

*Because not every surface age the same or deserves to be treated the same.*

All pavements wear, but not all pavements wear evenly. A shaded school walkway with foot traffic will age differently than a sun-drenched industrial loading bay. A permeable paver system near soft landscaping will behave differently than a sealed surface in a dense commercial plaza. And yet, too often, maintenance cycles are applied uniformly based on contract templates, not use-specific behaviour.

That's where failure starts; not in design or installation, but in the assumption that one plan fits all.

To design for real durability, we must map out intervention cycles that reflect actual usage conditions, not generalised categories. The question isn't just "what type of surface is this?" It's "how will this surface be used, abused, ignored, and maintained?"

Let's consider a few contrasts:

## 1. High-traffic commercial plazas

- **Surface risks:** Abrasion, UV exposure, organic staining from food or drinks
- **Cycle expectations:**
  - Cleaning: Weekly to biweekly, with occasional degreasing
  - Jointing inspection: Quarterly
  - Surface treatment (sealers or colour protectants): Every 3–5 years

## 2. Permeable paver car parks

- **Surface risks:** Joint clogging, sediment intrusion, weed growth
- **Cycle expectations:**
  - Vacuum sweeping or jet flushing: Twice yearly
  - Joint topping-up: Annually, based on site condition
  - Underdrain inspection: Every 3 years

## 3. Pedestrian linkways under tree canopies

- **Surface risks:** Algae growth, joint softening, uneven drying
- **Cycle expectations:**
  - Surface brushing or low-pressure washing: Monthly or after rain events
  - Algae-resistant coating reapplication: Every 2–3 years
  - Visual inspection of slip risk zones: Quarterly

## 4. Residential access paths or courtyards

- **Surface risks:** Light staining, occasional joint loss
- **Cycle expectations:**
  - Cleaning: Seasonal
  - Rejointing or spot-filling: Every 2–3 years
  - Structural reassessment: After 10 years or noticeable movement

These cycles aren't static; they evolve with climate, user behaviour, urban density, and even policy. In some cities, sustainability regulations now require documentation of stormwater performance, forcing owners to track permeability retention as a maintenance metric, not just a construction promise.

And in regions with shifting rainfall intensity or prolonged dry seasons, vegetation debris, dust accumulation, and runoff sediment loads will alter the frequency of intervention over time. The more precisely you define how a surface will live, the more accurately you can manage how it ages.

This is not just about prediction—it's more about thoughtful preparation. A surface designed with an unrealistic maintenance expectation becomes a liability. But when designed with honest cycles in mind and supported by proper material selection, detailing, and accessibility; the surface becomes a low-friction, long-life asset.

It's not about how little maintenance is needed. It's about how predictable, repeatable, and resource-efficient that maintenance becomes.

## Quantifying Whole-Life Cost: Beyond Initial Savings
*Because the cheapest surface is often the most expensive one to keep.*

Initial cost is easy to see. It fits neatly into tenders and budgets. It's a number with immediate impact. But long-term cost? That's quieter. Slower. Spread across years of repairs, interventions, cleanings, lost performance, and visual decline.

This is why whole-life costing is not a luxury consideration. It is the core measure of durability's true economic value. A surface that costs 15% more upfront but reduces joint failures, drainage problems, and major interventions over a 20-year cycle is not just technically better but also financially smarter.

Yet, in many projects, lifecycle cost is still treated as an afterthought, often replaced by a vague hope that maintenance will "take care of itself." It doesn't. And without a structured view of cost across time, clients end up paying for short-term wins with long-term losses.

## 1.  What Makes Up Whole-Life Cost?

To quantify real value, we must widen the frame. Whole-life cost includes:

- **Initial material and installation costs**
- **Maintenance cycles** – cleaning, joint refilling, algae removal, drainage checks
- **Repair costs** – labour, materials, and downtime for local failures or subsidence
- **Service disruptions** – blocked access, public complaints, safety claims
- **Resurfacing or full replacement costs** – especially when aesthetic or functional failure drives early decommissioning
- **Environmental costs** – water, chemicals, energy, carbon footprint of repeated interventions
- **End-of-life recovery or demolition expenses**

Each of these can be forecasted—not perfectly, but accurately enough to make comparative decisions between systems.

## 2.  A Mindset Shift: From Price to Value

Whole-life costing forces a shift from cost-cutting to cost-clarity. It moves the conversation beyond line items and into strategy:

- What's the break-even point between a low-cost system and a low-maintenance one?
- Where can operational budgets be offset by durability dividends?
- How does frequency of intervention impact carbon accounting or ESG reporting?
- Can design decisions be aligned with lifecycle-based procurement frameworks like ISO 15686, PAS 2080, or Envision?

These aren't theoretical frameworks; they're emerging procurement realities. And in many public-sector or sustainability-certified projects, whole-life costing is no longer optional. It's part of the tender score.

*"If you can't show how the surface saves over time,
you haven't finished designing it yet."*

## Embedding Durability into Asset Management Strategies
*Because a durable surface only matters if the system around it knows how to support it.*

The most well-engineered paving system can still underperform if the people responsible for managing it aren't equipped to maintain or measure its value. This is where the link between design and asset management becomes critical.

Durability doesn't stop at design. It needs to be tracked, scheduled, and embedded into the systems that govern maintenance, funding, and renewal. Without that connection, even the smartest surfaces are reduced to liabilities because no one knows when to intervene, how to do it properly, or what failure signs to monitor.

### 1. The Missing Layer in Asset Planning
In many municipalities and large estates, paved surfaces fall into an administrative blind spot. They're neither mechanical systems with sensors nor structural assets with routine audits. They're infrastructure by default that expected to last but rarely reviewed with data.

To embed durability into asset management, surface systems must be given the same treatment as lighting, HVAC, or roofing:

- *Scheduled inspections* based on performance thresholds
- *Preventive maintenance plans* aligned with design life and use environment

- *Lifecycle documentation* that includes specifications, intervention history, and cost profiles
- *Trigger thresholds* for repair, replacement, or reconditioning
- *Performance benchmarks* tied to service-level agreements (SLAs) or internal sustainability targets

When this framework is applied, pavement becomes more than a material layer. It becomes a measurable, managed system.

## 2. Aligning with ISO, LEED, and Envision Frameworks

This isn't just good practice—it's increasingly becoming a requirement. Standards like ISO 15686 (Service Life Planning) and PAS 2080 (for carbon in infrastructure) explicitly require durability integration into asset lifecycle models. LEED v4.1 and Envision ratings factor in long-term operational savings, maintenance inputs, and surface thermal behaviour—all of which trace back to paving system choices.

Even Environmental Product Declarations (EPDs) now include references to service life assumptions and end-of-life options, which asset managers must reconcile with actual wear patterns and maintenance history.

In short: durability is not just a design promise. It's an operational commitment.

## 3. The New Role of the Specifier

For designers and engineers, this means going a step further. Instead of simply handing over technical drawings, it involves:

- Providing maintenance playbooks tied to specific surface types
- Recommending inspection intervals and performance indicators
- Mapping high-risk zones based on stress and exposure conditions
- Anticipating failure modes and embedding pre-emptive strategies into O&M documents

It also means influencing procurement to work with clients to structure contracts and performance warranties that reflect total cost of ownership and not just initial delivery.


## Final Thought

A surface is only as valuable as the system prepared to take care of it. Too often, we mistake low maintenance for no maintenance. Or we celebrate a clean installation without considering what it will demand five, ten, or twenty years from now. But in truth, durability without a maintenance strategy is just a short delay on failure.

What we've seen in this section is that the real measure of paving performance isn't just how long it lasts, but how predictably, affordably, and sustainably it stays functional. A surface that costs less each year to manage, that avoids disruptive repairs, that keeps its visual integrity with minimal intervention—that's the surface that is working.

Whole-life value can't be seen on Day One. It's earned slowly, in the background, by how little attention the system demands, and how easy it is to respond when that attention is needed. Good paving disappears into the background of a well-functioning site. Great paving earns that invisibility by being quietly, consistently resilient.

Durability is not an end state; it's a relationship between design, use, and care. And when maintenance is planned from the beginning, that relationship becomes a partnership, not a rescue mission.

## 5.7. Lifecycle Assessment (LCA) and Longevity Credits

*Because sustainability isn't what a surface promises—it's what it proves over time.*

Sustainability goals are no longer guided by slogans; they're measured in numbers. For interlocking concrete pavers and other hardscape systems, LCA moves the conversation beyond "eco-friendly" materials and into quantifiable environmental impact: How much carbon did the product embody at production? How much energy does it demand in maintenance? How long will it last before replacement? What happens to it when it's removed?

These aren't abstract questions. They are central to how paving systems are now evaluated in green procurement, public infrastructure planning, and certification schemes like LEED, Envision, MyHIJAU, and BREEAM.

But here's the deeper opportunity: in a field where many materials may offer similar upfront emissions profiles, longevity becomes the differentiator. A surface that performs for 30 years with minimal intervention will almost always outperform one that needs to be replaced every 12 years—even if both used "green" materials to begin with.

***Longevity isn't just an operational asset—it's an environmental credit.***

This section explores how LCA frameworks apply to paving systems, how longevity is increasingly rewarded, and how specifiers and manufacturers alike can position durable pavements as a measurable contributor to environmental goals.

## Lifecycle Thinking for Hardscapes: Shifting from Product to System

*Because paving isn't just about what goes in the ground—but everything that happens before and after.*

For decades, the performance of paving systems was judged by compressive strength, aesthetics, or installation speed. Environmental considerations, if mentioned at all, were focused on recycled content or "low carbon" cement. But as infrastructure becomes increasingly tied to climate targets, durability and sustainability must be measured not in moments but in lifecycles.

This is where lifecycle thinking shifts the entire framework from evaluating a product at the point of purchase, to evaluating it across its entire journey:
- From raw material extraction
- To manufacturing and transportation
- Through installation, maintenance, and repair cycles
- And finally, to reuse, recycling, or disposal

In this framework, a paver is no longer just a concrete block; it's a carrier of environmental impact. And every stage it passes through either amplifies or reduces that impact.

### 1. LCA in Practice: What's Actually Measured?

Lifecycle Assessment (LCA) frameworks like **ISO 14040/44**, **EN 15804**, and **ILCD** calculate environmental impacts using data across four key phases:

- **Upstream (A1–A3):** Material extraction, transport, and manufacturing
- **Construction (A4–A5):** Delivery, site preparation, installation
- **Use and Maintenance (B1–B7):** Energy for cleaning, water for washing, frequency of jointing and repairs
- **End of Life (C1–C4):** Removal, transportation, disposal or recovery

The result is usually expressed in metrics such as:
- **Global Warming Potential (GWP)** in kg $CO_2$-equivalent
- **Energy Demand** (MJ/kg or MJ/m$^2$)
- **Water Use**
- **Ozone Depletion and Acidification Potentials**

This allows clients, regulators, and certifying bodies to compare different paving options, not just on specs, but on environmental consequence.


## 2. From Snapshot to Strategy: Why Lifecycle Thinking Changes the Conversation

LCA exposes a crucial truth: two surfaces can look identical, cost the same, and still behave entirely differently over time.

A conventional paver with minimal maintenance planning may require full replacement at year 12 due to jointing failure, surface degradation, or subsidence. In contrast, a more engineered system; using algae-resistant additives, stabilised joints, and modular repair design might remain serviceable for 30 years with only minor interventions. On paper, their initial environmental footprints might seem similar.

But when you account for:
- Two rounds of demolition and disposal
- Double the transport and reinstallation
- Repeated cleaning with high-pressure water or chemical treatments
- Material loss and landfill contribution

...it becomes clear: ***durability is environmental performance***.

And this is where system-level thinking matters. Because it's not just about choosing a "green" product. It's about designing an integrated system; where subbase, joint, restraint, and surface finish are selected to reduce long-term impact, not just initial cost.

## Longevity as a Sustainability Credit: What the Standards Are Starting to Recognise

*Because if a surface lasts twice as long, it shouldn't have to prove its value twice.*

For years, sustainable material certifications focused almost entirely on what something is made of mainly focusing on its recycled content, its toxicity profile, its embodied carbon at the point of manufacture. But now, that lens is widening. Increasingly, how long something lasts and how well it avoids replacement, intervention, or disposal is being recognised as a direct contributor to sustainability.

This is the emerging recognition of longevity as a credit-worthy asset. Green building frameworks like LEED v4.1, Envision, MyHIJAU, and BREEAM are beginning to formalize what engineers have known all along: a surface that stays functional with minimal inputs is more sustainable than one that claims low impact but requires frequent replacement.

### 1. How Longevity Enters the Credit Framework

Each certification system approaches longevity a bit differently, but the trend is clear:

- **LEED v4.1 (USGBC)**
  In the *Heat Island Reduction* and *Material Optimization* credits, LEED gives preference to durable, low-maintenance surfaces—especially those that maintain SRI values and structural integrity over extended lifespans.

  Surfaces with documented life expectancy, proven low-maintenance cycles, and modular repairability gain additional value during project scoring.

- **Envision (ISI)**
  Under *Credit NW 2.4: Maintainability and Longevity*, Envision explicitly rewards systems that minimize future replacement needs, reduce material consumption over time, and support adaptive reuse.

  Durable pavements that allow for non-destructive repair, modular replacement, or dismantling without landfill loss are seen as infrastructure assets—not just site finishes.

- **MyHIJAU (Malaysia)**
  MyHIJAU-certified products that include **Environmental Product Declarations (EPDs)** with durability data or reference service lives are better positioned in government projects, particularly in green procurement categories under JKR and local councils.

  Surfaces that demonstrate reduced lifecycle water usage, energy for cleaning, and material input per square metre over 10+ years are actively favoured.

- **BREEAM (UK)**
  In *Mat 05: Designing for Durability and Resilience*, BREEAM requires project teams to consider environmental conditions and product life when selecting external materials. Long-lived surfaces with protective features (like algae resistance, impact resistance, or UV stability) earn higher scores, especially when backed by performance testing and durability claims.

## 2. The Operational Upside

Beyond certification points, this recognition of longevity is reshaping procurement language, warranty expectations, and urban design policy. In high-visibility infrastructure such as transit corridors, institutional campuses, civic plazas; planners are now tasked with choosing materials not only for appearance or environmental footprint, but for how long they'll stay out of the repair schedule.

In practical terms, that means:

- Tenders requesting ***minimum design life validation***, not just material spec sheets
- Bonus weighting for products with ***modular reuse*** or ***EPD-backed service life claims***
- Demand for surface systems that contribute to ***reduced maintenance carbon*** over a 20–30 year model

The message is clear: longevity is no longer invisible. It's becoming a quantifiable environmental asset, just like recycled content or low VOC emissions.

*"The greenest surface isn't the one that starts clean. It's the one that stays working and keeps others away from having to touch it."*

## Documenting Durability in an LCA Context

*Because if it isn't measured, it isn't credited—and if it isn't credited, it won't be protected.*

Longevity may now be recognised in sustainability frameworks, but recognition only matters when it can be proven. And in Lifecycle Assessment (LCA), proving durability means turning technical performance into verifiable data; that can be modelled, validated, and compared.

Too many paving systems claim 20–30-year lifespans based on anecdotal observation or legacy assumptions. But when it comes to Environmental Product Declarations (EPDs), procurement assessments, or certification submissions, those claims carry weight only when backed by quantified durability documentation.

That's where LCA modelling becomes both a design tool and a proof mechanism.

## 1. What Does Documented Durability Actually Require?

To be integrated into LCA frameworks and sustainability assessments, durability must be **linked to measurable performance attributes and standardised references**. That includes:

- **Declared service life (DSL)** or **Reference Service Life (RSL)**, as defined in ISO 15686
- Evidence from **accelerated aging tests**, such as freeze–thaw cycling, abrasion resistance (EN 1338 Annex D), or UV exposure simulation
- **Jointing retention data**, algae resistance lab results, or permeability degradation over time (for permeable systems)
- Manufacturer-supplied **third-party test reports** and performance warranties
- **Historical maintenance data** where available, especially for public realm installations or municipal pilot projects
- Inclusion of lifespan assumptions within the product's **EPD (EN 15804 format)** or related environmental documentation

These inputs allow LCA tools to accurately model the environmental impact over a full lifecycle; including maintenance, repair, and end-of-life scenarios.

And critically, they allow specifiers, contractors, and clients to move beyond guesswork—making performance-based comparisons between systems with different costs, lifespans, and long-term behaviours.

## 2. The Role of EPDs and Service Life Claims

An Environmental Product Declaration (EPD) is not just a document; it's a gateway to participation in certified, green infrastructure. But an EPD that lacks service life justification, or defaults to overly generic assumptions (e.g., "50 years" without validation), risks being ignored in serious assessments.

Leading manufacturers now include:

- Declared lifespans based on testing
- Durability modifiers that allow LCAs to adjust embodied carbon per year of use
- Scenarios for modular reuse or secondary applications (e.g., crushed reuse in subbases)
- Maintenance frequency profiles to estimate resource consumption over time

In effect, they're turning durability into environmental credit—one dataset at a time.

### 3.  A Note on Transparency and Risk

Durability data also helps manage risk—for designers, for clients, and for regulators. When service life is based on evidence, not marketing, it builds trust. When it includes performance tolerances and end-of-life scenarios, it opens the door to better circularity planning, realistic maintenance budgeting, and carbon-informed decision-making.

*"Data doesn't just defend your material. It defends your decisions."*

## Designing for LCA Success: Practical Moves that Matter

For most designers and engineers, Lifecycle Assessment feels like something that happens later at downstream from the design table, managed by consultants, and built from assumptions. But in reality, LCA outcomes are shaped by decisions made long before the model is run.

What material is chosen.
What finish is specified.
How joints are detailed.
Whether the system drains.
Whether it can be repaired modularly or must be ripped out entirely.

These decisions are not just functional—they're environmental.

The power of lifecycle thinking is that it shifts sustainability from product marketing into design intelligence. A paving system that performs well in an LCA isn't one that relies on special labels; it's one that was intentionally crafted to reduce impact over time.

## 1.  Practical Design Moves That Influence LCA Outcomes

If the goal is a surface with lower carbon per square metre **per year of service**, here's where the real impact lives:

- **Prioritize long-lived systems.**
  Not just in theory—select materials and designs with documented service life of 25+ years and maintenance schedules that support it.
- **Design for modularity and localized repair.**
  If one unit fails, you should not have to replace the entire surface. Reusability and selective replacement reduce embodied carbon dramatically over time.
- **Control water and bio-load.**
  Algae-resistant coatings, proper slopes, and edge drainage reduces the cleaning frequency; thus, directly lowering water, chemical, and labour inputs.
- **Use stabilised jointing in erosion-prone zones.**
  Prevents premature washout, reducing the need for regular topping and reactive maintenance.
- **Limit excessive sealants or coatings.**
  These often wear quickly and add reapplication cycles to the LCA model without substantial durability gains.
- **Ensure compatibility with mechanical maintenance.**
  Surfaces that can be cleaned with energy-efficient sweepers or low-pressure water reduce operational footprint.
- **Specify EPD-backed products where possible.**
  Choose materials that provide declared and verified environmental data including end-of-life impact and service life assumptions.
- **Avoid overdesign.**
  Oversizing thickness or reinforcement "just in case" inflates embodied carbon with minimal benefit if it doesn't align with expected use loads.

## 2.  From Design to Decision-Making

More than ever, paving decisions are being evaluated by carbon per square metre per year. That's not a number on a product sheet; it's the outcome of many small, deliberate choices. The goal is not perfection. It's clarity. Clarity in how the surface behaves, how it wears, how often it's touched, and what happens when its first life ends.

***Designing for LCA success isn't about checking sustainability boxes. It's about designing a material system that ages well—economically, operationally, and environmentally.***

Because the best sustainability story isn't told at procurement. It's told by the surface itself, five, ten, and twenty years after it was laid.


## Final Thought

*Sustainability is not a label—it's a timeline.*

In the end, what makes a paving system sustainable isn't just what it's made of. It's what it becomes over time. How it performs. How often it's touched. How gracefully it ages. How easily it returns to the material cycle when its first life ends.

Lifecycle Assessment forces us to stop thinking in fragments. It replaces isolated performance metrics with long arcs of consequence. It asks us to track what happens *not just when we build*, but when we clean, when we repair, and when we try to start over.

And the answers it rewards are not always obvious. They favour systems that last; not by resisting change, but by being ready for it. They favour decisions made early about pattern, slope, drainage, joint, coating, binder—that ripple through decades of use. They reward what remains quiet: the surface no one complains about, because it works.

> ***"The lowest carbon paver may not be the one made greenest,***
> ***but the one that never had to be replaced."***

Durability, once seen as a bonus, is now environmental currency. It earns trust. It reduces emissions. And most importantly, it extends the life of everything we've invested in—materials, energy, labour, and time.

## 5.8 Chapter Reflection

**Durability isn't just what stays intact—it's what stays valuable.**

We often treat paving durability as a test of strength. But strength alone doesn't make a surface endure. What makes it endure is how well it absorbs pressure, adapts to change, and recovers from wear—not once, but over and over again.

In this chapter, we've moved beyond compressive strength charts and abrasion ratings. We've looked at real durability; the kind that lives in slope angles, joint widths, compaction layers, and maintenance logs. The kind that doesn't just resist failure but anticipates it. Designs for it and prevents it.

Durability is a discipline. It requires curiosity, humility, and a willingness to accept that surfaces don't live in the same world they're drawn in. They live in weather. In movement. In misuse. And most of all, in time.

It's also where sustainability gets real. Because a surface that fails early can't be green—no matter what it's made from. And a system that holds its value over decades, quietly doing its job with minimal intervention, becomes one of the most powerful environmental tools we have.

That is the heart of lifecycle performance. Not perfection. Not invincibility. But foresight. Systemic thinking: and the commitment to design a surface that lasts.

# Chapter 6: Circular Thinking in Hardscapes
*A New Lens on Surface Design*

We don't typically think of pavements as temporary. Once they're laid, we expect them to stay put—quiet, functional, and forgotten. But in an era where material flows, carbon budgets, and adaptive infrastructure define sustainable design, that expectation is due for a shift.

Because permanence is not the same as durability. And in fact, some of the most resilient systems are those that were never designed to be static.

Welcome to circular thinking in hardscapes: a design philosophy that sees interlocking surfaces not as inert layers to be buried and replaced, but as modular assets that can be lifted, reused, reimagined, and reintegrated into future applications. In this model, a paver isn't just a product; it's part of a system with lifecycle agility.

What makes this possible isn't just the material. It's the way we design the interfaces: the joints, the restraints, the base layers, the way pavers are held together *and* how they can come apart. In this view, the true sustainability of a paving system is measured not just by how long it lasts but how easily it lets go when change is needed.

> **"The most responsible surface isn't the one that stays forever.
> It's the one that knows how to leave with value intact."**

This chapter explores the technical strategies, design principles, and real-world practices that make circularity possible in interlocking concrete systems without compromising performance, safety, or aesthetics.

# 6.1 Designing for Disassembly and Reuse

*Because if it can't be taken apart without destruction, it was never truly modular.*

The beauty of interlocking pavers lies in their promise of flexibility. Yet too often, that promise ends with installation. Pavers are mortared into place. Joints are filled with rigid materials. Bedding layers are compacted with no separation from the subgrade. What began as a modular system becomes a monolith. True circularity begins at the interface. If a paver is to be reused; either on-site or elsewhere, it must be retrievable. And that means designing not just for strength or layout, but for clean, damage-free separation when the time comes.

## 1. Key Principles of Reversible Paver Design

- **Dry-laid, not wet-fixed**
  Avoid mortars, adhesives, or rigid surface sealants that bind units into one unliftable mass. Dry-laid pavers on well-compacted, separated base systems are inherently more recoverable.

- **Layer separation and geotextile inclusion**
  Using separation layers (such as geotextiles or polymer grids) between bedding and subbase prevents intermixing, allowing base materials to be reused or re-graded without contamination.

- **Accessible, maintainable edge restraints**
  Mechanical edge restraints that can be lifted or reset without excavation make removal viable and modular layout boundaries reusable.

- **Stabilised but non-permanent jointing**
  Polymer-modified or kiln-dried joint sand can offer stability during use but still be loosened for removal—if the right grain size and compaction method are used.

- **Uniformity in thickness and format**
  Reuse is only as easy as re-alignment. Standardised dimensions across batches, sites, or phases increase the likelihood of direct reuse without trimming or rejection.

## 2. Design Details That Enable Disassembly

- Leave "starter gaps" in large areas—buffer zones or demarcated rows that can be removed first to initiate lift-up access.
- Label or map high-stress zones vs low-stress zones—this can help future users salvage more material from less-worn regions.
- Use installation methods that enable *reverse sequencing*: what goes in last comes out first, avoiding cross-lock patterns that trap units permanently.

## 3. Why This Matters

Designing for disassembly is not just about environmental virtue. It has direct operational and economic value:

- Allows for **phased redevelopments** without full teardown
- Reduces **construction waste volume**
- Enables **redeployment** of surface materials across sites
- Preserves value in the asset—turning a surface from sunk cost into recoverable stock

In large campuses, industrial estates, or public infrastructure zones, this thinking opens new possibilities for adaptive hardscapes where surfaces evolve with usage patterns, masterplan changes, or policy shifts.

*"A truly modular system is not one that installs fast,*
*it's one that reinstalls just as easily."*

# 6.2 Demountable Urbanism: Applications and Opportunities

*Because not every surface is meant to last forever—and that's not a weakness. It's a design opportunity.*

Cities are no longer static. Streets are repurposed for events, plazas become markets, laneways double as performance spaces, and logistics zones evolve with new site demands. Yet, most surfaces are still designed as if they'll never change.

Demountable urbanism challenges that default. It asks: What if we designed for surfaces that were strong enough to perform but light enough to leave? What if durability didn't just mean longevity in place, but longevity through adaptability?

In this view, pavers become more than infrastructure. They become strategic tools—used, reused, lifted, relocated, and reassembled in response to shifting priorities. And the interlocking system, long celebrated for ease of installation, becomes the foundation for reversible, responsive public space.

## 1. Real-World Scenarios Where Demount-ability Creates Value

### a) Seasonal Plazas and Event Spaces

Public squares and pedestrian walkways in urban cores often serve dual purposes—foot traffic by day, market or festival venues by weekend. Demountable pavers allow surface-level services (e.g., pop-up stalls, cabling, drainage) to be accessed or adapted without excavation.

### b) Construction Staging and Phased Developments

On campuses, hospital expansions, or business parks, temporary access roads and equipment pads are needed during early phases. Demountable surfaces provide stable working ground yet can be lifted and reused as those spaces transition into landscape or light-traffic zones.

## c) Institutional Masterplans

Universities, R&D facilities, and industrial estates are increasingly built in phases. Surfaces installed during Phase 1 are often removed or reconfigured in Phase 3. With modular design, early-stage paving becomes a future resource, not waste.

## d) Logistics and Industrial Flex Zones

Loading areas, container yards, and warehousing aprons change shape and use pattern as operations evolve. Demountable hardscapes allow surfaces to grow, contract, or be reoriented in response to equipment upgrades or spatial optimization efforts.

## e) Community-led Development and Tactical Urbanism

In lower-income housing projects, informal settlements, or community gardens, modular pavements offer a rare combination: structural quality and non-permanence. This allows surfaces to be donated, shared, or reallocated between pilot zones and finalized spaces.

## 2. A Circular Asset, Not a Static Layer

Traditional paving systems are treated as endpoints in capital projects: once installed, they are owned and maintained until failure. Circular systems reframe this: the surface is not an endpoint but a node—a place where material can land, work, and move on.

This has profound implications for:

- **Carbon accounting** (reuse reduces embodied emissions)
- **Public infrastructure financing** (pavers become assets, not just expenses)
- **Disaster recovery** (temporary surfaces for housing or logistics can be repurposed quickly)
- **Urban innovation pilots** (testbed installations, pop-up zoning, tactical pedestrianisation)

### 3. Barriers to Adoption and the Opportunities They Reveal

Despite its value, demountable urbanism is still underused. Why?

- Procurement norms favour permanence: few tenders request temporary but durable surfacing.
- Designers assume reuse limits aesthetics: they forget that uniformity, not blandness, enables visual harmony.
- Maintenance teams aren't trained for lift/reinstall workflows: retraining is rarely factored in.
- Cost misconceptions: while upfront costs may be slightly higher, lifecycle savings and reusability more than compensate.

These barriers are real but they're not technical. They're systemic. And they reveal where the next frontier lies in training, contracts, specification language, and long-term urban planning.

## 6.3 Material Passports and Digital Traceability

*Because a paver that lasts 30 years is only useful if someone remembers where and what it is.*

We've spent decades designing surfaces to be durable. Now, we must make them knowable. Because in a circular economy, it's not enough for a material to be reusable—it must be recognised, validated, and trusted long after it was first installed.

This is the purpose of material passports: digital or physical records that capture everything about a product's identity; its source, composition, performance characteristics, and lifecycle history. When applied to paving systems, they become the connective tissue between design, construction, asset management, and reuse.

In short: a paver becomes more than a block of concrete. It becomes a traceable unit of value.

## 1. What Is a Material Passport?

A material passport is a data file or label that travels with a material across its lifecycle. For interlocking pavers, this could include:

- Batch and product ID
- Manufacturing date and location
- Compressive strength and abrasion class
- Mix design (e.g., SCM % or recycled content)
- Installation date and contractor ID
- Intended service life and expected maintenance intervals
- Digital twin record in BIM or asset platform

It's not just an inventory tool—it's a performance ledger that allows clients, contractors, or urban managers to make informed decisions when reuse, relocation, or recycling becomes an option.

*"Without traceability, every reused paver is a guess.*
*With it, every unit is a certified resource."*

## 2. How Traceability Is Being Implemented

Globally, several frameworks and technologies are emerging to support traceability in hardscape and modular construction materials:

- **QR Code & RFID Tagging**
  Embedding codes in edge units, pallets, or corner pavers allows quick field access to specs via smartphones or scanners (used in trials in developing countries).

- **BIM-Linked Material Databases**
  Systems like Autodesk Revit, Trimble, and open-source IFC models increasingly support material-level tagging, enabling pavers to be tracked as individual or grouped elements within infrastructure models.

- **Digital EPD Libraries & Blockchain Pilots**
  Tools like Madaster (Netherlands), EC3 (US), and the EU's Materials
  Passports Platform (MPP) offer product traceability, linked to
  verified EPDs, including carbon footprint and reuse guidelines.

- **MyHIJAU and CIDB Malaysia Potential**
  While not yet mainstream, Malaysia's green product certification
  ecosystem is well-positioned to include traceability as part of digital
  construction and GBI-linked procurement standards.

## 3. Why This Matters in Practice

Material passports and traceability:

- Enable safe reuse by retaining technical identity and performance
  classes
- Support carbon and circularity reporting in green procurement and
  LCA models
- Facilitate warranty validation across ownership transitions
- Reduce testing redundancy when reused across projects
- Build confidence in reused materials as reliable inputs—not
  downgrades or risks

This becomes especially important in:

- Municipal reuse schemes
- Campus-scale redevelopment plans
- Industrial zone asset management
- Large event infrastructure (e.g., Olympics, expo parks)
- ESG-reporting corporate campuses or government precincts

*"A reused paver with a passport isn't second hand—it's pre-certified."*

### 4. Challenges and Next Steps

While the concept is strong, adoption still faces barriers:

- Lack of standardisation across regions and manufacturers
- Cost concerns for tagging or database integration
- Fragmented responsibilities between designers, suppliers, and asset owners
- Limited policy mandates requiring traceability on public projects

But this is changing. With the EU's Digital Product Passport Regulation coming into force for construction products, and ISO 22057 supporting data templates for environmental and material documentation, global precedent is forming.

The role of the designer? Begin including traceability specifications—at least for public, large-scale, or circular pilot projects. Collaborate with suppliers who provide data-rich product specs and treat material identity as part of performance, and not just paperwork.

*"In the future, every material will come with a story.*
*What matters is whether we've written it down."*


# 6.4 Second-Life Pavers: Reuse, Regrade, or Reprocess

*Because when a surface reaches the end of its role—it shouldn't reach the end of its value.*

The most sustainable paving system isn't always the one that stays in place the longest. Sometimes it's the one that gets a second life. Or a third. Or that returns to the supply chain as raw input for a future surface. This is the logic of circularity in practice: not endless preservation, but intelligent transformation.

***Second-life pavers*** are those that have fulfilled their original use but retain enough structural or aesthetic integrity to be reused—either directly, downgraded, or reprocessed. In a world of growing material demand, shrinking landfill space, and rising sustainability requirements, they represent a largely untapped resource stream.

But tapping into that stream requires clarity: not every paver will be re-laid. Some may be crushed. Others regraded. Others still may find use in non-traffic applications. The key is to match post-use condition with next-best application; safely, efficiently, and with full material knowledge.

## 1. Reuse: Lift and Relay

This is the ideal case. Pavers are removed without damage, cleaned, and re-laid in a new location—either with the same performance expectations or in a reduced-stress environment.

Best for:
- Modular pavers with minimal wear
- Systems with documented history (see: material passports)
- Areas where colour or wear consistency is not critical (e.g., utility zones, community spaces)

Success factors:
- Consistent sizing (within 1–2 mm tolerance)
- Structural integrity intact (no corner breaks, major chipping)
- Proper jointing material to allow clean separation
- Lifted using manual or mechanical extractors, not demolition tools

## 2. Regrade: Downgrade and Reassign

If pavers have cosmetic fading, edge wear, or micro-chipping, they may no longer meet performance standards for commercial or high-load use. But they can be reassigned to lower-demand areas:

- Walkways and footpaths
- Garden borders or planters
- Temporary installations or experimental zones
- Rural road shoulders or edge details

Grading systems—either internal or based on EN 1338 tolerance categories can help sort pavers into "Class A" (full reuse), "Class B" (visual downgrade), and "Class C" (reprocess). This approach maximises utility without incurring the cost of full processing or testing.

*"A chipped corner doesn't end a paver's life.*
*It just changes where it belongs."*

## 3. Reprocess: Crush, Sort, and Reinvent

For pavers that cannot be reused due to severe wear, breakage, or outdated design, material recovery remains a valuable pathway.

Crushed concrete units can serve as:
- Recycled coarse aggregate in new concrete
- Subbase material for new paver beds or asphalt
- Fill for landscape grading or slope stabilisation
- Drainage layers beneath permeable systems

Key considerations:
- Ensure pavers are free from hazardous coatings or heavy metals
- Sort by type—pigmented vs plain, sealed vs unsealed
- Screen and wash to remove joint sand and fine contaminants
- Follow established gradation and absorption specs for structural reuse (e.g., BS 8500 or ASTM C33 guidance)

### 4. A Strategic Approach to Second Life

To integrate reuse into standard workflows, consider these tactics:

- Include second-life options in tender specs, offering credit for planned reuse, not just new installation
- Maintain inventory of removed pavers, classified by batch, condition, and storage location
- Collaborate with contractors and recycling centers for paver-specific crushing and screening
- Educate procurement teams: reused does not mean inferior—when managed correctly, it means resource-smart

And most importantly: create feedback loops. Every second-life deployment teaches something about installation durability, failure points, and modularity design. Those lessons are fuel for the next generation of circular systems.

*"Reusability isn't the end of a paver's story,*
*It's what allows the story to continue somewhere new."*


## 6.5 Chapter Reflection

*Circularity begins not with what we reuse, but how we choose to design in the first place.*

We often think of hardscape surfaces as permanent fixtures—laid once, serving silently, and eventually removed in fragments. But this chapter has shown us a different reality. One where pavers are not just surface material, but strategic infrastructure assets; flexible, traceable, and inherently reusable when designed that way.

True circularity in hardscaping doesn't arrive by accident. It must be planned, supported, and communicated. It demands technical detailing as much as it demands cultural shift. From edge restraints to batch tracking, from tender language to second-life logistics—every layer of the system must be ready for change.

And this shift isn't abstract. It's already happening—in seasonal plazas that lift and relay, in phased campuses that reassign surfaces, in digital models that store a material's story long after it's been walked on.

*"The real sustainability of a paving system is not only in how long it lasts, but in how long its value survives."*

We no longer live in cities that stay still. Our infrastructure shouldn't, either. With modularity, traceability, and reuse-ready design, interlocking pavers become more than concrete; they become carriers of resilience, economy, and intelligence. Circular thinking in hardscapes isn't the future. It's the next responsible step.

# Chapter 7: From Pavement to Purpose

*Integrating Principles, Driving Change, and Building a Resilient Surface Future*

We've spent hundreds of pages talking about surfaces. Their strength, their sustainability, their failures and redesigns. But this chapter isn't about surface at all. It's about what lies beneath—not in the subbase, but in our mindset.

Because when we choose a paving system, we're not just solving a technical challenge. We're expressing a worldview. One that either believes infrastructure is disposable or sees it as a long-term actor in the planet's future. One that defaults to permanence or plans for change. One that values what is easily built or what is quietly sustained.

Interlocking pavers, modular systems, reversible construction; these are not just materials. They are carriers of intent. Of foresight. Of humility. Of boldness. The choice to design a paving system that drains properly, that reflects heat, that can be reused and documented and disassembled—*that's not just a design move*. It's a signal.

> ***"A signal that we're no longer paving over the problem,***
> ***but paving toward the solution."***

As we close this book, we're not ending with more specifications. We're ending with a shift. From compliance to contribution. From surface to system. From footprint to future.

# 7.1 Reconnecting the System: What We've Really Been Building

*Because when we build pavements, we're not just shaping the ground—we're shaping behaviour, climate response, and cultural intent.*

We rarely pause to ask what a pavement *means*. It's just the thing beneath our feet—silent, solid, expected. But that expectation is precisely the issue. For too long, paving systems have been treated as background elements: passive, purely functional, separate from the urgent conversations happening elsewhere in sustainability, public health, resilience, and climate action.

But the reality is different. Pavement is climate infrastructure. Pavement is public safety. Pavement is policy, carbon, runoff, heat, cost, and adaptability. It frames how people move, how water flows, how cities breathe, and how environments age.

When we install a surface, we're making choices about:

- Whether water gets absorbed or redirected
- Whether heat is retained or reflected back into the air
- Whether a surface can be repaired or must be replaced
- Whether a material becomes waste—or enters a second life
- Whether maintenance will be routine—or crisis-driven
- Whether the system we build today will burden—or benefit—those who inherit it tomorrow

These are not trivial consequences. They're systems-level levers. And interlocking pavers especially when designed for sustainability, offer a rare convergence: a surface-level product that touches deep infrastructure systems.

## 1. Connecting to Environmental Systems

At its best, a permeable paver is not a luxury feature. It's a *hydrological intervention*. It reintroduces infiltration, slows runoff, filters pollutants, and relieves stormwater systems under pressure from intensifying rainfall and urban density.

High-SRI pavers? They're not just cooler to walk on—they're microclimate regulators. By reflecting solar radiation and emitting less absorbed heat, they reduce ambient temperatures, support thermal comfort in cities, and even lower building energy loads near the ground level.

Materials that are reusable, recyclable, or low carbon in their manufacture aren't just "eco." They're tools in *decarbonization*. They help reduce the embodied emissions footprint of our cities and help us meet the targets we so often miss.

These are systems impacts. They're climate-level impacts. But they start with something as ordinary as a 200x100 mm concrete block.

## 2. Connecting to Societal Systems

Paving also affects people. Not just by how it looks or wears, but by how it frames behaviour, accessibility, and safety.

- An algae-resistant path reduces fall risk for the elderly in shaded parks.
- A modular public square that can be reconfigured enables cultural events, civic engagement, and economic activity.
- A surface that reflects, rather than absorbs, heat contributes to lower rates of heat exhaustion; especially among vulnerable groups like children, outdoor workers, or the unhoused.
- A design that allows for clean disassembly reduces disruption during repairs and invites the community to **co-own public space** through tactical urbanism, planting beds, and event setups.

These are societal dividends from technical decisions. They only exist when surfaces are designed not in isolation, but as infrastructure with consequence.

*"The irony of paving is that it's seen as inert—when in reality, it might be the most interactive layer in the built environment."*

Because every step taken on a paver is part of a much larger movement: toward smarter cities, safer streets, cooler zones, cleaner runoff, and longer-lasting assets.


## 7.2 From Specification to Stewardship
*Because sustainability doesn't stop at the tender—it begins there.*

In most projects, the life of a paving system is defined by a document: the specification sheet. Dimensions, strength class, slip resistance, colour, layout pattern. Boxes ticked. Standards referenced. Approval granted. Done. But what happens next? After the install crew leaves, after the pavers settle, after the ribbon is cut—who ensures the system continues to function as it was intended?

This is the missing link in many sustainability conversations. We specify for compliance. But we rarely steward for performance, and that gap between what's designed and what's maintained is where most sustainability promises quietly to fall apart.

Because no matter how eco-friendly, permeable, or modular a paver is, its value depends on what happens over time:
- Was it maintained properly?
- Was it inspected before failure showed up?
- Was the jointing renewed before erosion compromised interlock?
- Was the algae resistance monitored, or simply assumed?
- Was the layout preserved during nearby construction, or damaged and patched poorly?

These are not just operational questions. They're questions of stewardship confirming the ability to carry design intent forward into meaningful, measurable performance.

## 1. The Lifecycle Gap

Most project lifecycles end at commissioning. But true sustainability lives in the **next phase**:

- Preventive maintenance planning
- Surface audits and condition rating systems
- Budgeting for service intervals based on use and exposure
- Keeping digital records (e.g., BIM, material passports) for replacement planning
- Coordinating with cleaning teams, not just installation crews
- Linking surface performance to asset valuation and public satisfaction metrics

This is not glamorous work. But it's where durability becomes real—not because we said the paver lasts 30 years, but because someone made sure it did.

## 2. The Shift: From Project-Based Thinking to Systemic Responsibility

For designers and engineers, this means expanding their role:

- Don't just specify a joint width—recommend a maintenance schedule for topping material.
- Don't just select an algae-resistant mix—brief the facilities team on how to inspect and validate it.
- Don't just choose high-SRI pavers—include instructions for gentle cleaning that won't degrade surface reflectance over time.
- Don't just hand over the layout plan—embed it in the asset registry for future repair reference.

For contractors, it means installing with reuse in mind, documenting processes, and treating the paver not as a disposable layer but as a potentially reconfigurable material stock.

For asset owners and municipalities, it means investing not only in procurement, but in preservation, training teams, tracking performance, budgeting for small interventions that avoid large disruptions.

And for suppliers and manufacturers, it means making information transparent: not just "green" marketing, but actual performance data, EPDs, wear profiles, compatibility guides, and circularity pathways.

*A paving system doesn't succeed when it's laid. It succeeds when it's still working, still safe, still relevant years later, with fewer inputs and less intervention.*

Stewardship is the bridge between what was specified and what is sustained. And if we're serious about environmental impact, long-term cost control, and infrastructure resilience, **that bridge must be designed as deliberately as the surface itself.**


## 7.3 The Role of the Professional: Designing with Consequence

*Because the choices we make in the design room echo across decades of use.*

Every paver that cracks prematurely, every surface that floods, every joint that unravels under foot traffic—these aren't just construction defects. They're signals. Signs that somewhere, a decision was made without full context. Or worse, without full care.

In the conversation around sustainability and resilience, we often focus on materials and systems. But behind every good or bad system, there's a professional. Someone who chose that mix. Approved that layout. Ignored that maintenance constraint. Or advocated for a smarter detail that the client didn't understand.

That means the burden and the opportunity rests squarely on us.

Because professionals don't just design surfaces. They design consequences. And in infrastructure, those consequences show up slowly: in maintenance records, in storm drain blockages, in cooling loads, in trip hazards, in budget overruns 8 years down the line.

*"Design is not an act of creation. It's an act of anticipation."*

## 1. Rethinking Professional Responsibility

In the age of climate urgency, rising urban complexity, and material constraint, the role of the paving professional must expand. It's not enough to:

- Specify to minimum code
- Follow precedent blindly
- Treat the surface as the architect's concern
- Hand off maintenance to "operations"
- Assume contractors will preserve reversibility
- Ignore lifecycle cost because it's "not in the scope"

Instead, the emerging role is one of integrative stewardship:
- Understanding how paving systems interact with drainage, lighting, greenery, urban heat, accessibility, and pedestrian flow
- Designing with material science and user behaviour in mind
- Speaking in LCA, EPD, and SRI metrics—not just shape and colour palettes
- Collaborating with contractors to design for disassembly, not just durability
- Documenting assumptions clearly so future users inherit intelligence, not guesswork

This is not added work. This is core value—and in a circular, low-carbon economy, it will become the baseline expectation of every competent practitioner.

## 2. Collaboration: The Missing Ingredient in Most Surface Failures

Paver failures rarely happen in isolation. They're the result of disconnects between:

- The architect's aesthetic vision
- The engineer's technical detailing
- The contractor's site improvisation
- The facility manager's capacity
- The budget holder's short-term thinking

When these roles don't align, surfaces fall apart—sometimes literally. But when they do align, something remarkable happens: the system becomes smarter than the sum of its parts.

- Maintenance teams know what to look for
- Clients understand the logic of long-term choices
- Contractors protect design intent through execution
- Designers feel empowered to specify for performance, not just appearance
- Everyone involved builds something that not only looks good but works *quietly* for decades

The new professionalism in paving design isn't about prestige. It's about precision, accountability, and humility.

It's the willingness to design a surface that most people will never notice; because it simply does its job, year after year, storm after storm, without complaint.

## 7.4 The New Benchmarks: Performance, Proof, and Purpose

*Because sustainability is no longer a claim—it's a commitment with evidence.*

Not long ago, paving systems were evaluated by a narrow set of metrics: compressive strength, abrasion resistance, slip coefficient. If the surface met the code and didn't fail on day one, it passed.

But those days are over. Today's benchmarks are broader, deeper, and more demanding because the stakes are higher.

Clients aren't just asking, "How strong is this?"

They're asking:
- *How hot will this get in July?*
- *How much runoff will it generate?*
- *How much energy did it take to cure?*
- *What happens when it needs to be removed?*
- *How often will we need to clean it?*
- *What's the embodied carbon per year of use—not just per square metre installed?*

In this context, success is no longer defined by the material alone but by its performance in context, its impact over time, and its alignment with purpose.

### 1. Benchmark 1: Performance That Lasts

This isn't just about initial strength. It's about **functional endurance** across dimensions:

- Surface integrity under dynamic loads
- Joint retention through thermal and water cycles
- Skid resistance over years of pedestrian and vehicular wear

- SRI retention under UV exposure and biological staining
- Algae resistance not in marketing, but in rainy-season performance audits
- Permeability that doesn't clog after the first monsoon

These aren't abstract metrics. They're observable, testable, and increasingly demanded in performance-based procurement frameworks.

## 2. Benchmark 2: Prove it, Not Just Promise

A paver can no longer say, "I'm sustainable." It has to show it. That means:

- Environmental Product Declarations (EPDs) aligned with ISO 14025 and EN 15804
- Lifecycle Assessment (LCA) models that factor in maintenance and reuse
- Real-world case studies, audits, or post-occupancy evaluations
- Durability testing under accelerated aging, freeze–thaw, and traffic simulation
- Albedo and emissivity verification under ASTM or ISO protocols
- Water infiltration and sediment retention rates documented over multiple seasons

The shift is clear: from *intent* to *evidence*. From soft claims to hard data.

And with emerging mandates for product carbon labelling, digital material passports, and green public procurement, this level of accountability is no longer optional—it's operational.

## 3. Benchmark 3: Purpose—A System That Serves Something Bigger

At the heart of it all, we must return to purpose. What does this surface *exist* to do?

Not just carry loads. But reduce carbon. Manage water. Protect users. Enable reuse. Reflect heat. Signal care. Surfaces can no longer be designed in isolation from their environmental and social roles. If they are, they fail—even if they technically perform.

> *"The purpose of a paving system isn't to be admired. It's to make everything around it work better for longer, with less."*

## 4. The Way Forward

In the next decade, expect to see:

- Public infrastructure tenders requiring LCA-backed durability claims
- Developers demanding maintainability modeling alongside aesthetic mockups
- Smart cities integrating paving systems into digital twins for asset tracking
- Material tags and QR codes built into edges—not for marketing, but for provenance
- Maintenance and repair metrics linked directly to carbon and water footprint calculations
- Surfaces that are valued less for what they cost at installation—and more for what they *prevent* over time

> *We are no longer in the era of "good enough."*
> *We are in the era of proof—or be replaced.*

And that's a good thing. Because it means every surface, we design from here on out must mean something and prove it, every day it stays in place.

## 7.5 Looking Ahead: Where Innovation Must Go Next

*Because what we've built so far isn't enough for the challenges we now face.*

For most of modern history, paving innovation has focused on durability and efficiency: stronger mixes, faster curing, easier installation. And while those gains matter, they're no longer enough. The pressure we face today; be it ecological, urban, or economic requires something far more radical:

Innovation that is ***responsive, regenerative***, and ***integrated*** into the larger systems shaping our cities and climate. Concrete surfaces are no longer static backdrops. They're data points, drainage tools, climate actors, and public health enablers. The next generation of paving design must reflect that complexity; not by being complicated, but by being *intelligent*. So where must we go from here?

### 1.  Digital Twins and Smart Surface Integration

The rise of digital twin cities demands that physical infrastructure speak the language of data. Pavers can no longer be passive. They must be part of live systems.

- Embedding sensors to monitor load, temperature, vibration, and infiltration
- Linking paving modules to building information models (BIM) and GIS platforms
- Tracking lifecycle use and wear data for predictive maintenance
- Using AI to simulate traffic impacts, heat retention, and permeability degradation over time

A paving system that knows when it needs cleaning or when it's nearing end-of-life is not a gimmick. It's risk mitigation. Cost control. Climate adaptation.

## 2. Carbon-Conscious Manufacturing and Circular Production Loops

Net-zero targets can no longer be met with wishful thinking. Innovation must:

- Lower embodied carbon per function, not just per tonne
- Prioritize non-fired, low-energy binders like LC3, geopolymers, and recycled fillers
- Shift to renewable-cured curing regimes and closed-loop water systems
- Design factories for resource circularity—with waste recovery, solar integration, and modular product output

We must begin to think of paver production not as an industrial process but as a regenerative system that aligns with planetary boundaries.

## 3. Hyper-Localized Design and Climate Adaptivity

There is no universal paving solution. The next frontier is contextual intelligence:

- Pavers designed specifically for high rainfall tropical zones vs. arid heat islands
- Joint systems that adjust for soil movement in expansive clay zones
- Algae-resistant microstructures tailored to monsoon drainage basins
- Textured finishes calibrated for pedestrian volume, not just visual design
- UV- and heat-resistant pigmentation that maintains high SRI over long-term exposure in equatorial climates

**"This is not "customization"—this is *resilience by design.*"**

## 4. End-of-Life Intelligence and Reverse Logistics

We've explored this already in Chapter 6—but here's the next step:

- Integrating **removal pathways** directly into design software
- Tagging batches with **reuse suitability scores**
- Creating shared **urban stockpiles** of second-life materials
- Embedding **recyclability incentives** into supplier contracts
- Partnering with *reverse logistics platforms* to bring circularity out of theory and into street-level practice

Because disassembly is not demolition, and reuse is not improvisation - It's engineering.

## 5. New Professional Training for Systemic Paving

Finally, the most powerful innovation may be cultural:

- Training civil engineers to design for maintenance, not just for strength
- Teaching architects to collaborate on edge restraint strategies
- Bringing landscape designers into SRI, permeability, and LCA conversations
- Developing multi-disciplinary pavement labs that combine material science, urban planning, and environmental design
- Updating university curriculums to include circular product thinking, not just design codes

*"The future isn't waiting for a better paver.*
*It's waiting for a better mindset across the board."*

We are now in a moment where the urgency of change meets the maturity of tools. What's needed is leadership; designers, developers, and institutions willing to stop doing what's familiar and start doing what's *consequential*.

*Innovation isn't always about something new. Sometimes it's about finally doing what we already know is possible—at scale, with integrity, and without delay.*

# Final Reflection: The Surface as a Signal

*Because the ground we build on reflects the values we build by.*

Every city tells a story; not just through its skyline or landmarks, but through its surfaces. The pavements, plazas, walkways, carparks, and corridors. The places most people never stop to admire. The ones we design not for attention, but for action. For durability. For passage. For purpose.

But what if we looked at those surfaces differently? What if every paver wasn't just concrete—but **a statement**?

That sustainability matters here.
That water is respected here.
That heat is mitigated here.
That change was anticipated here.
That maintenance won't be an afterthought here.
That future users were considered—before the first joint was laid.

Because the surface is not separate from the system. It is the interface. The layer between the built world and the living one. And every detail—every slope, every joint, every material choice—is a signal about how we see that relationship.

> *"A smooth surface that drains, reflects, resists algae, and can be reused isn't just functional—it's philosophical."*

It says: We care not just about how things start, but how they continue. Not just what we install, but what we leave behind.

And so, as we conclude this book, the goal is not to remember every technical specification or benchmark; but to remember this:

Sustainability is not a checklist.
Durability is not just strength.
Innovation is not just what's new.
And paving is never just paving.

It is the platform upon which resilience walks, economy moves, and climate decisions land—literally. It is a surface, yes. But it is also a message.

Let it be a message we're proud to stand on.

# Glossary

## Materials & Construction

**Abrasion Resistance** – The ability of a surface to withstand wear caused by friction from pedestrian or vehicular traffic.

**Aggregate (Coarse/Fine)** – Granular material (e.g., gravel, sand, recycled concrete) used in concrete mixes and subbase layers for strength and stability.

**Binder** – The cementitious material that binds aggregates together in concrete. Can include OPC, GGBS, fly ash, or LC3.

**Compressive Strength** – A material's capacity to withstand loads that tend to reduce size. Key structural metric for concrete.

**Hydraulic Conductivity** – A measure of how easily water passes through a material or soil—critical for subbase and drainage design.

**Joint Stabilizer** – A polymer or resin used to lock jointing sand in place, improving washout resistance and weed suppression.

**Open-Graded Aggregate** – A type of aggregate with little or no fine material, allowing for high void content and permeability.

**Paver Interlock** – The mechanical friction and contact developed between pavers, contributing to surface stability under load.

**Permeable Concrete** – A concrete mix designed with interconnected voids that allow water to pass directly through the paver body.

## Sustainability & Lifecycle

**Carbonation (of Concrete)** – A natural process where carbon dioxide reacts with hydrated cement, influencing long-term strength and $CO_2$ uptake.

**Cradle-to-Gate / Cradle-to-Grave / Cradle-to-Cradle** – Terms used in LCA to define the boundary of environmental impact assessments.

**Durability Index** – A multi-factor metric that considers structural, chemical, and environmental resistance over time.

**Embodied Energy** – The total energy required to extract, manufacture, and transport a material to the point of use.

**Green Infrastructure** – A planning approach that integrates vegetation, soil, and water systems into urban development for environmental benefit.

**Heat Island Effect** – The phenomenon where urban areas experience higher temperatures than surrounding regions due to heat-absorbing materials.

**Low-Impact Development (LID)** – Urban planning strategies aimed at managing stormwater as close to its source as possible, often using permeable surfaces.

**Operational Carbon** – The carbon emissions associated with maintaining and operating infrastructure during its use phase.

**Resilience** – The ability of infrastructure or ecosystems to withstand and adapt to changing climate, usage, or stress conditions.

**Sustainable Drainage System (SuDS)** – A water management strategy that mimics natural drainage to reduce runoff and promote infiltration.


## Urban Planning & Design Integration

**Bioswale** – A landscaped channel designed to slow, collect, and filter runoff using vegetation and soil.

**Civic Hardscape** – Publicly accessible paved environments like plazas, promenades, or pedestrian zones, which serve both utility and social function.

**Contextual Design** – An approach that adapts infrastructure to local climate, culture, and environmental conditions.

**Desire Line** – The path pedestrians naturally take across a surface—often not aligned with formal walkways.

**Multifunctional Surface** – A paving system that provides multiple benefits (e.g., drainage, cooling, navigation, reuse) beyond load-bearing.

**Tactical Urbanism** – Temporary, often community-led interventions in public spaces that test ideas before permanent infrastructure is installed.

**Transit-Oriented Development (TOD)** – A planning concept focused on maximizing access to public transport through dense, walkable development patterns.

## Testing, Standards & Certification

**ASTM / EN / ISO** – Acronyms for international standards bodies: American Society for Testing and Materials, European Norms, International Organization for Standardization.

**Environmental Product Declaration (EPD)** – A third-party verified document that quantifies a product's environmental impact throughout its lifecycle.

**Life Cycle Assessment (LCA)** – A method for evaluating the environmental burden of a product or system over its entire lifespan.

**MyHIJAU** – Malaysia's official green certification mark for environmentally preferable products and services.

**Solar Reflectance Index (SRI)** – A measure combining albedo and emissivity to express how hot a surface will become in the sun.

**Water-Sensitive Urban Design (WSUD)** – An approach that integrates the urban water cycle into built environments, often through permeable or vegetated surfaces.

# Sources & Further Reading

## Chapter 1: Why Surface Matters

- United Nations Environment Programme (UNEP). *Greening the Blue: Infrastructure for a Sustainable Future*. UNEP, 2021.
- World Bank. *What a Waste 2.0: A Global Snapshot of Solid Waste Management to 2050*. 2018.
- American Planning Association (APA). *Planning and Climate Change*. PAS Report 2011.
- U.S. EPA. *Urban Heat Island Basics*. https://www.epa.gov/heatislands

## Chapter 2: What Makes a Paver "Eco"?

- Interlocking Concrete Pavement Institute (ICPI). *Permeable Interlocking Concrete Pavement: Design, Specification, Construction, and Maintenance*. 2020.
- ASTM C1781 / C1781M-21. *Standard Test Method for Surface Infiltration Rate of Permeable Unit Pavement Systems*.
- DGNB System. *Criteria Set for Sustainable Site Use and Stormwater Management*. German Sustainable Building Council.
- LEED v4.1. *Heat Island Reduction Credit Requirements*. U.S. Green Building Council.
- DWA A 138 (Germany). *Planning, Construction and Maintenance of Facilities for the Percolation of Pre-treated Wastewater*. 2005.

## Chapter 3: Smart Design for Sustainable Function

- U.S. EPA. *Green Infrastructure Design and Implementation.* 2022.
- CIRIA (UK). *The SuDS Manual (C753).* 2015.
- Singapore PUB. *Code of Practice on Surface Water Drainage.* 2021.
- Malaysian Standards Department. *MSMA 2nd Edition: Urban Stormwater Management Manual for Malaysia.* DID, 2012.


## Chapter 4: Materials That Make a Difference

- IEA. *Tracking Cement.* https://www.iea.org/reports/cement
- MCI/NRMCA. *Recycled Concrete Aggregate: Applications and Standards.* 2022.
- BSI. *BS EN 197-1: Cement – Composition, Specifications and Conformity Criteria.*
- ASTM C494. *Standard Specification for Chemical Admixtures for Concrete.*
- Ellen MacArthur Foundation. *Circular Economy in Construction.* 2020.


## Chapter 5: Product Durability and Lifecycle Performance

- Concrete Society. *Durability Planning in Concrete Construction.* 2019.
- U.S. Federal Highway Administration (FHWA). *TechBrief: Long-Term Performance of Interlocking Concrete Pavements.* 2016.
- CEDR Transnational Research Programme. *Road Asset Management for Sustainable Development.* 2021.
- BRE Centre for Sustainable Products. *Whole Life Costing and Life Cycle Assessment for Sustainable Building Design.* 2020.

## Chapter 6: Circular Thinking in Hardscapes

- Ellen MacArthur Foundation. *Completing the Picture: How the Circular Economy Tackles Climate Change*. 2019.
- Circular Buildings Toolkit. *Guide to Materials Passports and Disassembly-Ready Construction*. 2022.
- ICPI. *Designing for Reuse: Guidelines for Modular Pavement Systems*. 2020.
- C40 Cities. *Circular Economy in Cities: A Roadmap for Action*. 2021.


## Chapter 7: From Pavement to Purpose

- IPCC. *Climate Change 2023: Mitigation of Climate Change*. Contribution of Working Group III.
- Royal Institute of Chartered Surveyors (RICS). *Whole Life Carbon Assessment for the Built Environment*. 2017.
- International Federation of Landscape Architects (IFLA). *Global Accord on Sustainable Infrastructure*. 2022.
- UN-Habitat. *Streets as Public Spaces and Drivers of Urban Prosperity*. 2013.

# About Author

**Arreshvhina** is an engineer with more than 16 years of experience across concrete technology, sustainable systems, and operational leadership, his career has been forged at the intersection of hands-on execution and long-range thinking. He holds a Master's degree in Civil Engineering, specializing in Concrete Technology, and an MBA that complements his technical insight with strategic acumen.

As General Manager for one of Malaysia's leading building materials groups, he oversees the operation of multiple interlocking concrete plants, where the science of strength meets the complexity of systems. His leadership spans R&D, manufacturing innovation, and organizational transformation—often under quiet pressure, always with intent.
But beyond plant floors and planning dashboards, **Arreshvhina** is driven by deeper questions:

What does it mean to build responsibly?
Where does material end and meaning begin?
How do we construct not just infrastructure, but understanding?

These questions form the foundation of **The Measured Mind**—a platform where engineering is treated not just as a profession, but as a philosophy. Here, he writes about the unspoken dynamics of construction, the overlooked energy behind durability, and the silent weight of decisions in the built world. His work bridges disciplines, dissolves silos, and invites dialogue between form and foresight. Through essays, reflections, and technical deep dives, Arreshvhina offers more than information—he offers clarity in complexity, and structure where others see only scale.

**"We don't just build with materials—we build with decisions. And the quiet ones often matter most."**

— *Arreshvhina*

# Author's Note and Disclaimer

This book reflects not just years of research and hands-on practice—but a lived perspective shaped over 17 years in concrete technology, materials design, and sustainable infrastructure systems.

I used advanced writing tools, including generative AI, to support the development of this manuscript — primarily for drafting assistance, restructuring, and language refinement. However, the **technical content, frameworks, insights, and editorial decisions are entirely my own**. Every concept has been filtered through real-world engineering judgment, industry standards, and critical reasoning.

This book is not a sales document, nor a catalogue. It is a systems-level guide intended to challenge assumptions, deepen technical understanding, and elevate how we think about hardscapes in a climate-conscious world.

If something resonates here, I'm glad. If something provokes debate, even better. Because sustainable infrastructure is not a checklist — it's a conversation.

While every effort has been made to ensure accuracy, industry practices and regulatory frameworks evolve. Readers are encouraged to consult current standards, project specifications, and qualified professionals for site-specific decisions.

Ultimately, the ideas here are mine — and so is the responsibility.

- Arreshvhina -